New Technology
and
Industrial Relations

Warwick Studies in Industrial Relations
General Editors: G. S. Bain, R. Hyman and K. Sisson

Also available in this series

The Management of Collective Bargaining
Keith Sisson

Managing the Factory
P. K. Edwards

Conflict at Work
P. K. Edwards

Dismissed: A Study of Unfair Dismissal and the Industrial Tribunal System
Linda Dickens, Michael Jones, Brian Weekes and Moira Hart

Consent and Efficiency
Eric Batstone, Anthony Ferner and Michael Terry

Unions on the Board
Eric Batstone, Anthony Ferner and Michael Terry

The Changing Contours of British Industrial Relations
William Brown (ed.)

Profiles of Union Growth
George Sayers Bain and Robert Price

The Social Organization of Strikes
Eric Batstone, Ian Boraston and Stephen Frenkel

Shop Stewards in Action
Eric Batstone, Ian Boraston and Stephen Frenkel

Trade Unionism under Collective Bargaining
Hugh Armstrong Clegg

New Technology
and
Industrial Relations

Edited by
RICHARD HYMAN and
WOLFGANG STREECK

Basil Blackwell

First published 1988

Basil Blackwell Ltd
108 Cowley Road, Oxford, OX4 1JF, UK

Basil Blackwell Inc.
432 Park Avenue South, Suite 1503
New York, NY 10016, USA

British Library Cataloguing in Publication Data
New technology and industrial relations.
——(Warwick studies in industrial
relations)
1. Industrial relations——Effects of
technological innovations.
I. Hyman, Richard   II. Streeck, Wolfgang
III. Series
331   HD6971

ISBN 0–631–15982–7

Library of Congress Cataloging in Publication Data
New technology and industrial relations / edited by Richard Hyman and
   Wolfgang Streeck.
      p. cm.—(Warwick studies in industrial relations)
   Rev. and edited papers from international colloquium at Warwick in
June 1986.
   Includes bibliographies and index.
   ISBN 0–631–15982–7 : $39.95 (U.S.)
   1. Industrial relations. 2. Technological innovations.
3. Skilled labor. 4. Trade-unions. 5. Labor supply-Effect of
technological innovations on.   I.Hyman, Richard. II. Streeck.
Wolfgang. 1946–   . III. Series.
HD6971.N38   1988
331—dc19   87–29363

Typeset in 10 on 11.5 pt Times
by Photo·graphics, Honiton, Devon
Printed in Great Britain

Contents

Part III Skills, Deskilling and Labour Market Power

Part IV Trade Union Strategies

Part V Technological Innovation and Workplace Relations

Contributors

Peter Armstrong, Principal Research Fellow, Industrial Relations Research Unit, University of Warwick

Greg Bamber, Director of Research, Durham University Business School

Sabine Erbès-Seguin, Research Director, Groupe de Sociologie du Travail, Université Paris VII

Stephen Frenkel, Senior Lecturer in Industrial Relations, University of New South Wales

Jon Gulowsen, Researcher, Work Research Unit, Oslo

Beat Hotz-Hart, Research Director, Institut für Orts-, Regional- und Landesplanung, Zürich

Richard Hyman, Professor of Industrial Relations, University of Warwick

Otto Jacobi, Principal Research Fellow, Institut für Sozialforschung, Frankfurt

Harry Katz, Associate Professor, New York State School of Industrial and Labor Relations, Cornell University

Pertti Koistinen, Senior Research Fellow, University of Joensuu

Kari Lilja, Associate Professor of Business Economics, University of Vaasa

Reinhard Lund, Professor of Organisational Sociology, Aalborg Universitetscenter

John MacInnes, Research Fellow, Centre for Research in Industrial Democracy and Participation, University of Glasgow

Serafino Negrelli, Research Professor, Dipartimento di Scienze dell'Uomo, Università di Trieste

Robert Price, Lecturer in Industrial Relations, University of Warwick

Helen Rainbird, Research Fellow, Institute for Employment Research, University of Warwick

Jørgen Rasmussen, Associate Professor, Aalborg Universitetscenter

Arndt Sorge, Senior Research Fellow, International Institute of Management, Wissenschaftszentrum Berlin

Wolfgang Streeck, Professor of Sociology and Industrial Relations, University of Wisconsin–Madison

Michèle Tallard, Researcher, CNRS/IRIS Travail & Société, Université de Paris–Dauphine

Stephen Wood, Lecturer in Industrial Relations, London School of Economics and Political Science

Series
Editors' Foreword

The University of Warwick is the major centre in Britain for the study of industrial relations. Its first undergraduates were admitted in 1965. The teaching of industrial relations began a year later in the School of Industrial and Business Studies, which now has one of the country's largest graduate programmes in the subject. Warwick became a national centre for research in industrial relations in 1970 when the Social Science Research Council (now the Economic and Social Research Council) located its Industrial Relations Research Unit at the University. In 1984 the Unit was reconstituted as an ESRC Designated Research Centre within the School of Industrial and Business Studies.

The series of Warwick Studies in Industrial Relations was launched in 1972 as the main vehicle for the publication of the results of the Unit's research projects. It is also intended to publish the research findings of staff teaching industrial relations in the University as well as the work of graduate students. The first six titles were published by Heinemann Educational Books of London; subsequent volumes have been published by Basil Blackwell of Oxford.

While most previous Warwick Studies have been the products of individual authors or research teams, the multi-authored monograph also has an important role in disseminating our work. *New Technology and Industrial Relations* falls within this category; but in two respects it breaks new ground. It is the first study to derive from a conference organized under the auspices of the IRRU; and it is the first to involve a Visiting Professor whose presence in the Unit was made possible by a generous grant from the Leverhulme Trust.

During his stay at Warwick, Wolfgang Streeck collaborated with Richard Hyman in arranging an international colloquium, and the most important papers have been revised for publication in this monograph.

They provide a wide range of detailed evidence on the relationship between technical change and industrial relations, framed by a number of original and in some cases provocative suggestions for comprehending current trends theoretically. Though the contributors embrace a variety of contrasting perspectives, there is a striking level of agreement on one point: that there is no evidence of simple technological determinism. New technology may have major repercussions on industrial relations, but these are mediated by the policies and strategies adopted by employers, unions and governments; while some existing options may be foreclosed, others are opened. The varied ways in which social choices are adopted – or evaded – appears as a central theme of comparative analysis.

George Bain
Richard Hyman
Keith Sisson

Editors' Introduction

Technology, market dynamics and the texture of social relations and institutions are intimately connected: on this simple proposition there can be little argument. What is far more contentious is the nature of the interrelationship, or indeed the primary direction of any causal linkages. Competitive pressures may in some circumstances encourage, in others inhibit firms in decisions to invest in microprocessor technology. Conversely, such decisions affect their ability to compete in particular markets, and cumulatively may help determine the fate of whole sectors of a national economy. The pace and content of technological change are necessarily affected by the institutional framework of industrial relations: systems of labour law, the initiatives of government economic agencies, the strength, structure and orientation of trade unions. Reciprocally, transformations in production technology have repercussions on such institutions.

Where so complex an interplay of social forces is under investigation, the task of research and scholarship is more centrally one of interpretation and evaluation than of simple description. Problems of definition aside, it is a relatively straightforward matter to identify, say, which car manufacturers have installed robotic welding lines, and to specify trends in production and employment since their introduction. It is a different issue altogether to establish how far changes in jobs or sales are *attributable* to technical innovation, or to demonstrate the manner in which these changes themselves have been conditioned by government policies or bargaining relations with trade unions.

In addressing such essentially qualitative judgements, the consideration of cross-national experience is of enormous value. Many of the arguments and propositions concerning technical change stem from evidence drawn from a single national context. To generalize from what may be national peculiarities is clearly hazardous. A broader perspective can help establish what patterns reflect country-specific idiosyncrasies, and what form part of a more universal set of relationships.

Such considerations explain the genesis of this book. Comparative analysis is an important aspect of the work of the ESRC Industrial Relations Research Unit at Warwick, where one of the editors teaches and the other was Leverhulme Visiting Professor of European Industrial Relations during 1985. In the summer of 1986 we jointly organized a colloquium on 'Trade Unions, New Technology and Industrial Democracy', with financial assistance from the ESRC, and under the auspices of the European Group for Organizational Studies (EGOS). Most of the papers submitted, revised in the light of the conference discussion, are included in this volume. They offer a wide-ranging international survey: half of the authors come from seven separate countries in Continental Europe, another two from Australia and the USA. Together they provide a particularly detailed and extensive picture of the relationship between technical change and industrial relations in contemporary Western economies.

The chapters in this volume are organized within five distinct sections. Part 1 addresses issues of theoretical interpretation, methodology and research strategy. Sorge and Streeck begin by disputing the common assumption that the interrelationship of technical change and industrial relations can usefully be analysed without examining such factors as company strategies, work organization, vocational training and other societally specific institutions. They argue that the linkages among such factors are far more complex than is conventionally assumed – and than trade unions, as 'actors' in industrial relations, recognize in their policies.

This premise leads Sorge and Streeck to challenge deterministic views of technical change as leading necessarily either to deskilling or to upskilling of the workforce. Rather, the patterns of work roles associated with microelectronic technologies reflect *strategic choices* by company managements. They argue that among these choices – which are influenced by distinct national contexts – may be a new type of work organization which combines specialization with elements of a craft tradition: an option which can prove particularly attractive in the context of heterogeneous and unstable product markets. Nor is the causal relationship unilinear: the high skill level of the workforce in the West German car industry, they argue as an example, has encouraged employer strategies oriented towards diversified and flexible, rather than pure mass production, and quality rather than price competition. The distinctive feature of the application of microelectronics in production is precisely that it opens up such options, allowing (though not necessarily resulting in) a movement away from the Fordist model of standardized mass production and the associated rigid subdivision of labour.

How can such arguments be related to the conventional concerns of industrial relations? Sorge and Streeck suggest that the production strategies associated with Taylor and Ford, with their separation between conception and execution, tend to be associated with 'low-trust' worker responses (to use a term proposed by Alan Fox) and adversarial industrial relations. More specifically, employers may unwittingly foster an introspective occupational trade unionism committed to the defence of traditional job demarcations. Conversely, a system of broader and more flexible skills encourages a philosophy of industrial unionism with a less rigid approach to job definition and task reorganization.

In conclusion, Sorge and Streeck point to the very different orientations of management and trade unions towards the technology/industrial relations interface. Employers start with a primary concern with product and market strategies; production technology is selected in the light of these considerations; only then do labour questions attract systematic attention. Trade unions, by contrast, are centrally preoccupied with terms and conditions of employment; only intermittently do they address (let alone seek to influence) decisions on technical innovation; scarcely ever are they concerned with the overall thrust of corporate policy. In the main, therefore, unions can achieve a significant impact on technology only when buoyant markets allow them to exert 'resistance to change'; where economic conditions are less favourable unions are virtually impotent. Paradoxically, an unfavourable economic environment – by eroding the traditional segregation of production from industrial relations decisions – forces trade unions to address fundamental issues of company strategy if they are to defend their members' day-to-day conditions.

Hyman focuses on the same issue of a production system in flux, but from a very different standpoint. His primary concern is to indicate weaknesses in 'optimistic' assessments of the social consequences of technological innovation: interpretations encapsulated in the notion of 'flexible specialization' (a term which Sorge and Streeck replace by 'diversified quality production').

Four main criticisms are developed. Firstly, those who discern new socio-technical models displacing traditional Fordist principles tend to exaggerate the coherence and consistency of management strategy. Managements are typically faced with contradictory pressures and constraints, to which they respond with opportunistic and unstable policies. Secondly, the literature on 'flexible specialization' involves generalization from a narrow empirical base; much of the evidence is drawn from a limited range of manufacturing industries (in particular motor vehicles), whereas the major growth in contemporary employment is outside manufacturing altogether.

Thirdly, it is suggested that the assumption that new structures of work organization will engender harmony in place of adversarial relations must be regarded sceptically. Most crucially, the traditional craft worker derived his power (almost all such workers were crafts*men*) from access to the *external* labour market; by contrast, the restructuring of internal labour markets creates a new relationship of dependence on a single employer, and hence new dimensions of vulnerability. Such changes create in turn the potential for new forms of conflict.

Fourthly, it is shown that the broader social consequences of the 'flexible firm' involve more complex issues. One obvious feature is the sharpening of divisions within the labour force as a whole: a polarization between the relatively secure and advantaged and the increasingly vulnerable and exploited. Segmentation in the wider labour market connects to new segmentations within consumption; flexible specialization as the basis for 'bespoke' production rests on an ideology of consumer inequality through the promise of exclusive, superior products. For all affected by these developments – advantaged and disadvantaged alike – a key consequence is the subordination of human relations to the impoverishing dominance of commodity relations. Perhaps significantly, the concluding theme of Hyman's chapter parallels that of Sorge and Streeck: the need for trade unions to develop more ambitious concerns, transcending the traditional agenda of collective bargaining.

Hotz-Hart also focuses on managerial choice between neo-Fordist mass production methods and the more flexible and innovative pursuit of specialist market niches. His theme, based on detailed research in Britain, West Germany and the USA, is the way this choice is influenced by industrial relations institutions.

Two main forms of this influence are examined. The first occurs through the structure and functioning of labour markets. The second is the institutional patterning of the nature and extent of the costs borne by management in achieving change. Hotz-Hart argues that German employers are faced by significant institutional constraints which oblige them to negotiate seriously in order to introduce change; but that once agreement is reached, other obstacles are of little significance. American managements are affected by few constraints either from industrial relations institutions or from labour market structure. In Britain, by contrast, both forms of restraint are important.

Management strategies are explored in the light of these two contextual factors. Again, the 'pecularities of the English' emerge: the context encourages managements to consolidate existing product and market strategies, rather than to pursue radical innovations. Hotz-Hart concludes by arguing that while institutional structures condition technological development, they are themselves subject to historical evolution. Management attempts to alter or evade institutions which constrain market-oriented choices – in the process challenging the existing status of trade unions – thus represent one element in a historical dialectic of economic and social innovation.

At a totally different level of analysis, Lund and Rasmussen develop an interactionist analytical framework with affinities to that of Hotz-Hart. Their empirical concern is with the role of local and regional government in Denmark in encouraging product and production innovation. The conceptual centrepiece is the notion of interorganizational networks. Lund and Rasmussen specify methods for defining such networks, and go on to distinguish as areas of research the content of interorganizational relations, the environmental influences, and interorganizational dynamics.

Technological innovation in Denmark has occurred against the background of a long tradition of centralized collective bargaining. In the 1980s, comprehensive technology agreements have been negotiated between the national trade union and employer confederations. However, recent trends have shown that transformations in industrial location and occupational structure have consequences which transcend the traditional terrain of industrial relations. Unions are thus forced to respond to a blurring of the former demarcation between economic (trade union) and political (party) concerns.

An important feature of Lund and Rasmussen's study is their exploration of the differential effectiveness of local union representatives in formulating objectives and choosing appropriate strategies for their attainment. Thus the study provides significant insights for the analysis of the micropolitics of institutional adaptation to technical change.

The second group of chapters is concerned with management initiatives which link technical change to work reorganization. In the first and most general chapter, Negrelli begins by presenting the contrast between Fordist and flexible organizations – the topic already discussed in the first two chapters. It is perhaps significant that Negrelli writes from Italy, often seen as the paradigm of the growth of 'high-technology cottage industry'. How, he asks, does the employer gain workers' enthusiasm and commitment which are essential if the firm is to respond rapidly to unpredictable changes in the economic and technical environment?

His thesis is that such needs lead inevitably to the transformation of the traditional adversarial pattern of industrial relations. From the attempt to define a stable body of rules, bargaining becomes a continuous process of problem-solving, encouraging the development of systems of worker participation. International experience, Negrelli suggests, indicates that co-operative or 'corporatist' patterns of industrial relations do indeed accompany radical technological innovation in countries with a tradition of integrated collective bargaining at company level. In the absence of such traditions – notably in Italy itself – participative forms of regulation are inhibited. Yet here too there are significant recent examples of innovation in industrial relations institutions.

Against this background, Negrelli suggests that employers pursuing technical and organizational change must choose between four strategies which either adapt or overturn traditional bargaining arrangements. One approach is to reassert managerial prerogatives, marginalizing trade unions or displacing them altogether. A second strategy is in practice only slightly different: retaining the formalities of collective bargaining, but only on condition that unions totally accept management policies. (The development of 'concession bargaining' in the USA in the 1980s is an obvious example.) One innovative method which excludes trade unions is the cultivation of employee participation arrangements which are internal to the company and controlled by management. The fourth variant also involves establishing an internal machinery of representation and consultation, but in conjunction with the involvement of trade unions in joint problem-solving. Negrelli concludes that only the first and last alternatives provide a stable basis for management strategy.

The notion of a transition from rigid, authoritarian, Fordist work organization and labour relations to more flexible, participative structures as a consequence of microelectronic technology is addressed critically in Wood's study of the American car industry. He argues that the polarized stereotypes often applied in the literature are misleading; recent managerial strategies display elements from both models, and may reverse some features of Taylorism while reinforcing others. One reason is that the relationship between technology and work organization on the one hand, and industrial relations on the other, is contingent, depending on specific managerial policies. Such innovations as 'quality of working life' (QWL) schemes may be pursued (as Negrelli suggests) either with *or* without the cultivation of collaborative relations with trade unions; and the creation of employee participation mechanisms may, but need not necessarily, accompany the relaxation of Fordist production techniques.

Such ambiguities are explored in a detailed investigation which examines in particular employee participation and changing industrial relations in General Motors and Ford. In some respects these developments may be seen as the 'Japanization' of the American motor industry – a trend accentuated by the entry of Japanese manufacturers themselves, in partnership with their local competitors, as American-based car producers. In the Japanese model, 'employee participation' is directly and deliberately linked to systematic employer efforts to minimize unit labour costs. The model's corollary (at least in popular imagery) of enhanced job security for the workforce is not, however, generally replicated in the American experiments of the 1980s.

Wood goes on to contrast theories of 'flexible specialization' which connect advanced production methods to specialized and localized market niches, with the earlier 'world car' principle, whereby increased simplification and standardization were regarded as the only basis for survival when competitive pressures intensified. Again he considers the dichotomy misleading. The flexibility of new technology, he argues, is itself exaggerated: what is typically involved is the production of many detailed variants of a single underlying design. For the leading motor companies, transnational product and marketing strategies remain crucial. One central implication, though, is clear: new production technologies create an array of possibilities for managerial strategy in industrial relations; the outcomes are thus indeterminate.

The chapter by MacInnes examines management strategy in a major Scottish bank. He notes the traditional gender division within banking employment: the majority of the workforce consists of women in clerical and analogous positions, but men almost completely monopolize the professional and managerial 'career' grades. Despite the recent adoption of notional equal opportunity policies, women are still in practice discouraged from pursuing the promotion route.

Microelectronic technology creates the potential for a dramatic transformation in banking work: further downgrading many routine functions, creating a new career framework for computer experts, and in aggregate reducing employment opportunities. Decisions on the application of new equipment and new work organization are little influenced by collective bargaining, which is heavily concentrated at the national level while a vacuum of negotiating machinery and trade union representation exists at the workplace. MacInnes gives the specific example of the introduction of automatic telling machines, and the employment of new part-time female staff to operate them; this occurred by unilateral management decision without the involvement of trade union representatives.

One reason why management proved able to enforce its own policies is that the workforce saw little reason to resist. The earlier introduction of computerization (when competitive pressures were less intense) had few effects which employees considered harmful; their attitude to new technology was therefore relatively complacent. Moreover, technical innovation was seen as a means of eliminating boring aspects of current jobs. Thus despite strenuous trade union attempts to increase workers' awareness of potential long-term dangers, and demands for negotiation rights over the introduction of new technology, it was impossible to win the active support of the members themselves. Here too, management utilized its autonomy to pursue policies which combined elements of both Taylorism and 'flexible specialization'.

The third section of the book addresses the debate over the implications of technical change for skills. In English-speaking countries, the terms of the controversy during the past decade have been shaped by Harry Braverman's *Labor and Monopoly Capital*, where the degradation of work through the divorce of conception from execution is posited as the dominant characteristic of 'progress' within capitalist production.

In the 1980s, the refutation of Braverman has become a major academic preoccupation. Armstrong seeks to defend him against what would seem to be ill-founded criticism. There has been no serious challenge, Armstrong suggests, to major elements of Braverman's thesis: that in the early stages of capitalist industrialization the purpose of technical innovation was to undermine the market power of skilled workers; and that a similar objective was explicitly embraced by F. W. Taylor in his advocacy of 'scientific management' in the early twentieth century. Some of the empirical studies of the introduction of microelectronic technology suggest a parallel rationale. How then should research which leads to quite different conclusions be evaluated?

Armstrong emphasizes the qualified and conditional manner in which Braverman propounded his 'deskilling' thesis. His concern was with general, long-term tendencies; propositions developed at this level of analysis are not contradicted by the evidence of individual, short-term case studies. Moreover Braverman's definition of skill was stringent, involving the synthesis of conception and execution; most 'refutations' employ a weaker, 'degraded' conception of skill. Much of the evidence deployed by critics is derived from the transitional implementation phases of new technology, rather than from its operation in a stable state. Much also depends on the argument that technological innovation is not consciously and deliberately pursued by management in order to undermine workers' skills and job control. But Armstrong stresses that the actual motives of management are not the focus of Braverman's argument; to contradict his thesis, it would be necessary to demonstrate that managerial initiatives were *systematically incompatible* with a deskilling outcome.

What, though, are the *consequences* of changes in worker skills? Or if skill is defined less strictly than by Braverman, what are the social implications of different types of skill? Gulowsen's chapter, with its focus on power relations between workers (and their unions) and employers, is centrally concerned with such questions. Unlike Armstrong he is not concerned with general tendencies, but with internal differentiation within the working class. A classification of skills is suggested, based on two dimensions: general versus company-specific, and socially versus technically defined. Combining these dimensions gives four categories: traditional craftsmen, 'worker bureaucrats', 'company specialists', and the newly skilled.

Gulowsen explores the significance of different types of skill for workers' power. The traditional craft worker is relatively independent of the employer and well placed to organize collectively. The worker bureaucrats, with socially defined and company-specific skills, are also disposed to unionize, but dependence on a single employer is a potential source of weakness. Company specialists, whose skills are technically based but firm-specific, are vulnerable to changes in technology and work organization, and often find unionization more difficult than the first two groups. Newly skilled groups, such as electronic data processing technicians, possess the strength of independence from a single employer (at least while demand for their expertise exceeds supply); but the very openness of their market position encourages individualism rather than collectivism.

In evaluating current trends, Gulowsen points to the displacement of stable, socially defined skills by more fluid, technically based skills. Such tendencies are reinforced by the pressures of international product competition, which undermines the foundations of trade union strategies which aim to defend established skills within individual economies. Accordingly, those categories of skill which once supported the strongest trade unionism are in decline; new categories of skilled work offer unions a less amenable prospect.

Rainbird provides a further perspective of technical change, skills and unions. Her starting point is that trade unions accept the economic inevitability of technological innovation; but they are concerned to influence the manner of its implementation and its impact on workers' conditions and security. As a corollary of this concern, they naturally express an interest in the training available to employees affected by change.

Changing technology and work organization, she emphasizes, has contrasting consequences for different groups of workers: some acquire new and enhanced skills, others find their jobs deskilled, yet others are unemployed. Rainbird notes the varied meanings of the concept of skill, and stresses that the technical aptitudes required in a specific job may not be socially recognized (particularly in the case of female workers).

There is a paradox, she argues, in the application of new technology and new forms of work organization: many workers are required to display initiative and adaptability, so as to respond 'flexibly' to changing production requirements; yet at the same time the technical content of the tasks performed may be reduced and degraded. This creates an imprecision in the social definition of skills, providing a terrain for trade union intervention and struggle. In the current economic climate, unions are ill placed to win recognition for the often substantial expertise that workers without formally sanctioned skills are required to display within their restructured tasks. Conversely, traditional craft workers are often threatened by changes which both constrict their

former wide-ranging competences and challenge established job demarcations. For such workers, the predominant implication of flexibility
is the erosion of craft status. In the case of a minority (notably
electronic crafts), however, technological innovation creates the basis
for new and often higher skills. This may well lead to the particularistic
pursuit of new sectional interests.

Developments in trade union strategy are the subject of the fourth
part of the book. The opening chapter by Jacobi counterposes
'pessimistic' and 'optimistic' visions of the social impact of rapid
technical change. Against the apparent trends of the past decade, he
insists that there is a serious probability of long-term economic
expansion, a renewal of state economic intervention, new scope for
trade union initiatives, and at least the potential for an enhanced trade
union role in social policy-making.

Jacobi locates the current decade within a model of long waves or
Kondratieff cycles with a periodicity of 50-60 years. The recession of
the 1970s and early 1980s can be viewed as the final throes of the
cycle based on the new industries of the inter-war era, he suggests; an
expansionary phase stimulated by such innovations as microelectronics,
materials technology and genetic engineering – a 'new technological
paradigm' – is likely to ensue.

He concedes that many uncertainties are associated with this
optimistic prognosis. The creative application of new technology is a
hazardous undertaking; since individual capitalists may refuse to assume
the risks, state sponsorship or support is essential. The transformation
which Jacobi envisages will impose costs as well as benefits, both within
and between nations; the equitable mediation of such consequences is
again an essentially political problem and solutions will necessarily prove
contentious. Despite the opposition of conservatives, an interventionist
state will be even more necessary in the coming decades than in the
past. The effective line of argument, suggests Jacobi, will therefore be
not between interventionists and abstentionists, but between advocates
of a socially concerned, labour-oriented politics and one which favours
and embraces capital.

Because of the current strength of conservatism, trade unions face
a short-term crisis of survival. But if they maintain their organizational
integrity, Jacobi concludes, they will enjoy new opportunities as
economic circumstances improve and as new layers of workers perceive
grievances requiring a collective solution. An effective union strategy
will require sensitivity and sympathy towards the distinctive needs of
categories of the working class outside the traditional core of trade
unionism (male, full-time, relatively secure industrial employees). To
choose such an approach would require the rejection of the option
which some unions evidently find attractive: concentration on the

sectional interests of a privileged elite within an increasingly divided workforce.

Like several other authors, Bamber seeks to link current trade union policies on technology with earlier historical experience. He begins with a brief discussion of initial working-class responses to industrialization, noting the early nineteenth-century machine-breaking struggles of the Luddites. Trade unionists of subsequent generations have normally been most anxious to avoid accusations of Luddism; and Bamber argues that they were indeed increasingly committed to the orderly regulation of the effects of technical change, without directly opposing innovation.

Thus he asserts a strong continuity between unionism yesterday and today, when the almost universal policy is to accept new technology and seek merely to negotiate over terms. Bamber indicates a number of aspects of current policy: procedurally, the demand for a right to be involved in management decisions on technical change; substantively, claims for protection against job loss, training in new skills, the sharing of the benefits of higher productivity through increased pay. The construction of a coherent overall programme from the array of possible demands is itself a difficult problem of internal union politics, and is not always successfully achieved.

The following chapter contrasts two approaches apparent within North American trade unionism. Katz labels these 'co-operatist' and 'militant', and argues that despite some convergence in actual practice the two ideal-types do represent a key contemporary division.

Co-operatists are receptive to a 'problem-solving' model of collective bargaining which – in the face of severe competitive challenges to the employer – may point to the surrender of previous gains in wages and conditions. Conversely, militants reject the whole philosophy of concession bargaining; they consider that it leads to a competitive undercutting of union standards, and in the long run and in the aggregate does not achieve the desired result of saving jobs. Similar differences are revealed in attitudes to work reorganization and participation schemes: for co-operatists they are a welcome means of enhancing employee status and influence; for militants they are spurious and dangerous.

Despite their diametrical opposition on issues of company-level policy, co-operatists and militants display surprisingly few differences in their broader political perspectives; both endorse similar demands for increased government economic intervention, including protectionist measures in the case of vulnerable industries. This suggests that their ideological divisions are confined to an assessment of the scope for 'positive-sum' labour–management relations within the company. Attitudes to this question reflect contrasting assessments of the balance of labour market power (in part conditioned by the product markets of particular employers) rather than broader images of economic structure and class relations.

Katz concludes that both approaches contain major weaknesses. Co-operatists have no principled basis which would allow them to draw the line between concession and capitulation. On the other hand, militants have no rational basis for determining when to resist and when to make judicious compromises; the concessions which they do indeed agree thus seem arbitrary and reactive. Both wings of the labour movement therefore need a longer-term and broader vision to give coherence to their strategies.

Frenkel concentrates his attention on a single union, but one which has exerted considerable influence on the policies of Australian trade unionism as a whole: the Metalworkers' Union. It was prominent in the early 1970s, under favourable labour market conditions, in encouraging the development of shop steward organization to spearhead pressure for decentralized improvements in wages and conditions. In the past decade, however, Australian engineering has been particularly badly hit by recession and government deflationary measures, and the union has thus been obliged to seek new strategies.

The new approach has contained a major political dimension, involving an agreed programme with the Australian Labor Party (in government since 1983). Initially this principally involved the trade-off familiar in post-war European social democracy: trade unions would exercise restraint in bargaining and in particular hold back their more favourably placed sections; the government would guarantee the maintenance of real wage levels and would improve social benefits.

Increasingly, however, the union has been forced to develop policies which address changes within the workplace. In the 1980s engineering employers have extensively pursued rationalization measures, based not primarily on new technology but on other forms of work reorganization – for example, the introduction of Japanese-inspired 'just-in-time' schemes. There has also been a widespread introduction of employee involvement schemes which the union viewed as an attempt to bypass collective bargaining and union representation. While rationalization led to job losses, such schemes posed a threat to the status of the union itself.

Its response has centred around the theme of industrial democracy. Frenkel explores a variety of initiatives, which may be summarized under three headings. The first involves pressure for an explicit trade union role in industrial planning processes, particularly where companies receive government funds. The second is the attempt to negotiate company-level agreements which provide for union participation (particularly through shop stewards) in the detailed decision-making on production reorganization. Thirdly, a trade union pension fund has been established, partly in order to channel investment along lines compatible with union policy. Since all these initiatives are very recent, it is difficult as yet to assess their effectiveness.

The final section of this volume examines developments within the workplace. Four themes interconnect: management initiatives, worker attitudes, union strategies and the legal framework of interest representation.

Price begins by noting the ambitious objectives formulated by British unions in the 1970s: the negotiation of comprehensive technology agreements covering a wide agenda of procedural and substantive issues. The evidence is clear: few specific technology agreements have been achieved, and those that have fall far short of what was originally envisaged. Not only have unions failed to gain control over the choice of technology, job design and work organization; there have been few if any significant attempts to place such matters on the bargaining agenda.

Does this mean that unions are therefore irrelevant in the implementation of technical change in Britain? No, argues Price; judged against more modest criteria, their record is not one of failure. Though they have not succeeded in influencing the choice of technology (or even made the attempt to do so), unions have sought to modify its application, and have often been able to soften its impact on the workforce.

On the basis of a series of case studies, Price suggests that the influence of workplace trade unionism is often considerable. A key factor, he suggests, is the nature of existing relationships between management and union. Where co-operative industrial relations have prevailed, management is often willing to make concessions in order to sustain consensus. Thus union representatives have been able to negotiate over the detail of new socio-technical arrangements, at times obtaining significant alterations to management's original plans. So gains have been achieved – though typically as the 'positive-sum' outcome of a process imposing no significant costs on the employer. By contrast, where conflict and antagonism had previously characterized industrial relations, the unions were unable to exert effective influence over new technology. Negotiations were formal and perfunctory, and had virtually no impact on management's implementation process.

The theme of co-operative implementation of new technology recurs in the account of innovation in Finnish engineering. Koistinen and Lilja point out that in a traditionally rural society which industrialized very rapidly in the mid-twentieth century, technical change is accepted as a fact of life. Until recently, change was associated with economic expansion and hence regarded as a source of material advantage. Moreover the combination of technical change with industrial growth created a labour market in which considerable choice and mobility existed. The resulting individualistic orientations which this encouraged have not been eliminated by the onset of recession; on the contrary, increased competition for jobs has reinforced individualism and made workers ready to embrace further technological innovation.

Kostinen and Lilja report the findings of a study of three plants in which advanced CAD/CAM systems were introduced. To operate the new production sections, management deliberately selected young semiskilled workers with an interest in electronics. These workers generally felt that their skill and status had been enhanced: and because they now interacted in their work with technical engineering staff they enjoyed greater autonomy from supervisors. Older employees – from whom the workplace union leadership was drawn – tended to regard the changed methods of work organization with greater ambivalence. Often they felt their own skills had been devalued; yet they also saw modernization of the firm as a source of enhanced earnings and greater job security.

The co-operative implementation of technical change in engineering is contrasted with experience in the pulp and paper industry, where an isolated workforce has traditionally shown an unusual degree of militancy. Here, exceptionally, a national agreement has been signed which requires companies to negotiate with the local union representatives before introducing any major change in work organization. This has provided the basis for hard bargaining; but it has been used primarily as a lever to extract wage concessions, rather than as a means to alter the proposed work organization and manning levels on new technology. Overall, the authors demonstrate a significant dualism in Finnish industrial relations: there is a high level of strike activity over wage issues, but militancy is not carried over into attitudes to work reorganization and even to redundancies.

In the following chapter, Erbès-Seguin discusses the changes in French workplace representation which stem from legislation enacted by the 1981-6 Left government. In France the traditional pattern of industrial relations has involved national (or sometimes company) bargaining over wages and conditions of employment, but a minimal trade union role at the workplace. Managements have been vigilant in resisting any threat to their authority, and unions have been unable to challenge the power of the *patron*. In general, then, wage claims have served as a substitute for qualitative demands.

This separation of the workplace from the processes of collective bargaining became increasingly anomalous as the period of economic expansion came to an end, unemployment rose, and employers responded to competitive pressures with rationalization and work reorganization. The government was anxious to encourage technological and organizational innovation, but also to ensure the consensual implementation of change. This helps explain the legislation of the early 1980s, which for the first time imposed on employers an obligation (though limited) to bargain with unions, and also created a new institution – the 'expression group' – through which employees may from time to time voice opinions on questions of work organization and working conditions.

Erbès-Seguin offers an assessment of the experience of these new mechanisms. One clear conclusion is that the outcome has been very uneven. While many expression groups were established rapidly – usually through some process of workplace negotiation – coverage is still very incomplete. It is also evident that many of the new bodies have little more than a paper existence. In some cases they have provided a genuine measure of employee influence over issues of work organization, including the introduction of new technology. Against the tradition of authoritarian management, this is a significant development. Where workplace union organization is relatively strong, it has often been possible to develop effective collaboration between the new machinery and traditional bodies such as works committees and health and safety committees. Elsewhere, however, expression groups seem to be totally controlled by management, acting like employer-initiated quality circles. The pattern is thus confused, and leaves trade unions with continuing problems in mobilizing workers in the face of the challenges of the 1980s.

In the final chapter, Tallard compares the links between technological innovation and industrial relations in France and West Germany, paying particular attention to the impact of the law. In Germany there exist legally constituted works councils with significant powers but bound by a 'peace obligation'. Formally, trade unions have only a minor role in the workplace, but the social consequences of work rationalization can be addressed in collective bargaining (which occurs at multi-employer level). Trade union influence in France is weaker than in Germany, partly because of the division between competing ideologically based unions. Labour law is also far less extensive and intrusive, although – as Erbès-Seguin documents – its impact has now increased significantly. In this respect, suggests Tallard, France may be seen as moving towards the German pattern.

While employers in both countries are now subject to a legal obligation to consult employee representatives before introducing any major technical change, in neither case, argues Tallard, do these provisions have much practical effect. German employers can exploit ambiguities in the legislation, and labour courts have in a number of cases adopted a narrow definition of the powers of works councils in respect of new technology. French legislation confers few real powers, and the courts tend to accept the employers' interpretation.

Where statutory provisions prove inadequate, how far does collective bargaining provide workers with a more effective influence? In both countries, trade unions have attempted to control the implementation of technological innovation through negotiation with employers, singly or collectively. German unions have developed ambitious programmes of job protection and other safeguards for their members, which have encountered determined employer resistance. This has been reflected in a number of major strikes and lockouts in recent years, resulting in various agreements on 'protection against rationalization' and 'humanization of working life'. French unions have developed less systematic or extensive strategies for bargaining over new technology. A number of recent agreements with employers' associations and individual companies have included provisions on this issue, but their content has usually been weak.

Inevitably, the coverage of a symposium such as this is highly selective. Nevertheless, we are confident that the collection offers many original and provocative contributions to analysis and interpretation, as well as an important array of findings from empirical research. May this book stimulate further advances in comparative research.

Part I

Theoretical and Methodological Issues

1

Industrial Relations and Technical Change: The Case for an Extended Perspective

Arndt Sorge and Wolfgang Streeck

Introduction

The relationship between technical change and industrial relations has for too long been analysed in a deterministic, unidirectional and ahistorical framework, separated from the wider institutional and economic context in which it is embedded. This, we maintain, reflects a tendency of mainstream industrial relations theory to take the historically contingent institutional arrangement of a differentiated, specialized industrial relations system as the unquestioned basis of concept formation and theory building. As a consequence, the relationship between technical change and industrial relations came to be primarily viewed from the perspective of worker and trade union resistance to change, and the explicit or implicit analytical objective became to compare and evaluate different industrial relations institutions by the extent to which they either permitted technical change (the liberal wing of the discipline) or prevented it (the radical position).

Our central criticism of the traditional approach to the relationship between industrial relations and technical change is that it treats the latter as an exogenous factor that operates on the former from the outside – the inside being essentially conceived in terms of a semi-autonomous social subsystem in the pluralist-functionalist mould (Dunlop, 1958; Kerr *et al.*, 1960). As a result, the direction and substance of technical change appear to be independent of the social institutions that regulate the exchange between capital and labour. Industrial relations, both in theory and in practice, become limited to the *implementation* of (managerial strategies of) technical change and

appear to have nothing to do with its *conception*. Radical critics have
seen this as reflecting, and in fact reinforcing, the exclusion of labour
in capitalist societies from important industrial decisions, such as those
on technology. But this overlooks the fact that trade unions themselves
have often been quite content to limit themselves to negotiating with
industrial relations managers on wages and conditions, and dealing
with technology only in so far as it affects the latter. It also overlooks
the possibility that, while technical change may not be a direct subject
of industrial relations, it may nevertheless be influenced by it, and
while such influence may be indirect and latent, it may nevertheless
be important.

To assess fully the mutual relationship between industrial relations
and technical change, it is necessary to extend the scope of the inquiry
beyond the historically contingent configuration of differentiated
institutions of joint regulation that emerged in the post-war period and
that has come to be equated with industrial relations (Batstone *et al.*,
1984: 5). One possible gain may be an improved capacity of industrial
relations actors, and trade unions in particular, to take into account
the impact of their behaviour on technical change. Moreover, by
making strategic use of hitherto latent causes, trade unions may be
able to extend their range of activities and objectives beyond the
narrow limitations placed on them in the pluralist-functionalist model.
In fact, we believe that it will become increasingly essential for trade
unions to do precisely this at a time in which the relative sovereignty
and autonomy of the traditional industrial relations system are being
progressively eroded under the impact of rapid economic, organizational
and political change.

This chapter will look in particular at two clusters of contextual
variables that affect technical change and industrial relations both
separately and interactively. One describes what we call the organization
of work and skills, the other, the market and product strategy selected
by a given firm or industry. In the following sections, we will first look
at technical change and discuss the way in which it is linked to the
two clusters of variables. Next, industrial relations will be introduced
and analysed in their relationship to technical change as mediated
through the organization of work on the one hand and the strategic
selection of markets and product ranges on the other. In the Conclusion
we will try to draw out a number of possible consequences of our
argument for industrial relations and trade unions in the present period
of economic and institutional restructuring.

Technical Change and the Organization of Work and Skills

By organization of work and work skills we mean an array of variables
that are salient for industrial sociology, the labour process debate and

the sociology of management and organization. They include:

the degree to which execution or direct production is separated from planning, engineering and maintenance (the *functional* division, or specialization, of labour);

the extent to which management and supervisory functions are differentiated from indirect and direct production (the *hierarchical* division of labour);

the concentration of knowledge, expertise and specialized experience in certain positions and functions (planning, engineering, design, development, management), accompanied by the deskilling of direct production or subordinate jobs, which has also become known as the polarization of skills;

the numerical growth of organizational sub-units in management, engineering and planning;

the degree of rigidity which is inherent in the forms of division of labour mentioned above.

These variables can be divided into *qualificational* factors, or skills, and *organizational* factors which refer to characteristics of workflow organization or organization structure. We use the word qualification in the Continental sense, i.e. denoting the skills and knowledge embedded in a person or required for the achievement of a work task – whereas often in English usage, qualification refers to paper qualifications, diplomas or other more formal sources of proof that are more or less tenuously connected with a work role or personal knowledge and experience.

Our second block of variables is technology or technical change. There have been numerous studies on examples of technical change, and standard definitions of a variable purporting to represent technology have been attempted. But research experience has shown that standard operational definitions of such concepts are hard to obtain (see for instance Woodward, 1970). The concept of technical change has been developed largely in the context of the automation discussion; an earlier survey, for example, was entitled 'Technology, Technical Change and Automation' (Parker *et al.*, 1972, ch. 9). Yet it has been shown that we are not in fact dealing with a homogeneous phenomenon, i.e. highly correlated dimensions. Even at a fairly high level of standardization of variables, Child and Mansfield (1972) have demonstrated that a technology scale ought to be divided into separate measures of *production continuity* and *workflow integration* which are conceptually distinct and in both studies they only shared about 25 per cent of variance in common (1972: 376).

A further complication is that technical change cannot be considered as simply consisting of variation along a known and well-defined dimension. Whereas in the classic Blau and Schoenherr (1971) study, the degree of automation was measured by the existence or non-existence of a central computer in a labour agency, any such operational definition has become meaningless by now. New technology raises the need for new standard definitions of technical variables, making previous standard definitions obsolete as innovation occurs.

At present, the notion of technical change is largely connected with the application of microelectronics in products and processes. But arguably, the consequences of process applications need not be the same as those of product applications, and they may not be homogeneous in themselves. This chapter explores the consequences of *process* applications of microelectronics, i.e. in the form of measurement and control engineering, communication and information technology equipment. Paradoxically, changes in process engineering seem to be more conveniently related to changes in product markets than changes in product technology. While our argument focuses on process applications of microelectronics within the manufacturing industries, it may be possible to extend it to cover services. We also expect a more valid and refined argument from such a transfer, but we are not able to achieve it at this stage.

Process applications of microelectronics – i.e. CNC machines, CAD/CAM equipment, industrial robots, computer-aided engineering (CAE) and computer-aided production control (Gunn, 1982) – are today in the forefront of industrial sociology, and organization and management research. The procedure has, more or less, been to neglect the question of how to fit technical change into a standard scale. Researchers have selected particular cases of technology or technical innovation for study, and general conclusions have then been drawn by aggregating and discussing findings from individual cases (Manpower Services Commission, 1985; Braun and Senker, 1982; Gensior, 1986).

In the sociological tradition, the interaction between the organization of work, the generation and distribution of skills, and technical change has mainly been conceptualized using the notion of the social division of labour. After Durkheim (1964), the key trend identified for societal development and technical change has been an increasing division of labour. In industrial sociology, this trend has been interpreted in basically three different ways:

The first view may be called the *degradation of work* or *polarization of skills* approach. It was exemplified by Braverman (1974) and Kern and Schumann (1970), and held that in capitalist production, complex work roles are continuously broken up and divided in two ways: on the one hand into lower grade, more routine, simple and monotonous roles within a more segmented organization of workflow; and, on the other, into more demanding, responsible and varied roles founded on more elaborate education and training. Leaving aside a few conceptual controversies and empirically deviant cases – which, one might suggest, can always be found with respect to highly general propositions – this approach has proven highly pertinent through much of the post-war economic boom period and across many studies (Parker *et al.*, 1972: 118-19; Sorge, 1979). Yet it has also been apparent that the low-grade skills which emerged at the bottom of the industrial and services workforce hierarchies were particularly exposed to rationalization measures through technical and organizational change. In a long-term perspective the result may thus have been a numerical decline of degraded places of work.

This perspective thus shades into another one which may be called *decline and rise of skills*. This has held that after the degradation process, the evolution of work roles was in the direction of upgrading and enrichment. This reversal of the previous trend was seen to be due to the increasing prevalence of continuous-process production after the heyday of mass production. We have put the idea into terms defined by Woodward (1965), and she had also used her typology as an evolutionary scheme, but the evolutionary concept dates back to such authors as Friedmann (1950) and Blauner (1964).

Whichever interpretation was adopted, scholars have for long searched for a deterministic view of technical change and work that was able to explain a host of diverse incidents of change by referring to a universal long-range tendency. At the same time, there has always been an approach which stressed the importance of *socio-technical choice*. It was notably founded on research at the Tavistock Institute, and asserts that the evolution of work was in no way determined by the course of technical change but by the rationale or strategy adopted by leading decision-makers in an organization. According to this view, the optimal development of the technical, the social and the sentient systems of an organization was founded on a strategy of enriching skills and achieving an overlapping, rather than divisive, organization of work tasks.

With research on applications of microelectronics, the last view has gained increasing ground (see, for example, Sorge *et al.*, 1983; Kern and Schumann, 1984; Trist, 1981). Proponents of the polarization of skills thesis often turned more than before to a perspective admitting greater choice between alternatives. It was also often suggested that enrichment of skills was a consequence of 'new technology'. We would view this latter interpretation as an unfortunate variant of the original argument as it seems to maintain the technical determinism inherited from earlier research and public discussion. Our view is that there is, and has always been, some degree of choice. It may also be that the cost–benefit calculations of choices have increasingly been loaded towards a more organic (Burns and Stalker, 1961) organization with more enriched skills and overlapping work roles, which is in fact compatible with increased division of labour in the sense of an increase in the number of specialized occupations. But it may be that work roles at the bottom are less fragmented and the division of labour features a greater amount of overlap of tasks and skills. In this respect, it is conceptually most important to distinguish the *extent* and the *mode* of the division of labour. The first may be represented by the number of different specialisms (jobs, work roles, occupations) whereas the second is defined by the rigidity with which specialisms are separated – or, conversely, by the amount of overlap between differentiated sets of activities and skill and knowledge profiles. Organic organization should be defined on the basis of overlap rather than number of specialisms.

This distinction is particularly applicable to long-term as opposed to short-term changes. It is, of course, true that socio-technical choice always comprises functional equivalents, of less divided and specialized (craft) organization, on the one hand, and more divisive and specialized (bureaucratic) organization on the other (Stinchcombe, 1959). Such choice exists at any given moment. But these may be compatible rather than exclusive alternatives (Heydebrand, 1973). In the evolution of work and technology over time, the alternative to increasing specialization may not usefully be construed as decreasing specialization, i.e. as craft *instead of* specialization. In the long run, the more promising alternative seems to be specialization *with* a greater craft element. This does imply a broadening range of specialisms; however, these are less rigidly separated, but linked through organizational overlap, training or career trajectories.

Looking at the rationale of alternative strategies of organization and skilling, the advantages of a polarizing regime lie in the economies of scale arising from the production of larger batches for larger and increasingly homogeneous markets, uniform mass markets at the extreme. In such a regime, it pays to concentrate and develop separately a wide range of engineering, planning and preparatory tasks. This makes for a more rigid division of labour including degradation of direct work. This logic, which has economic and organizational angles, was clearly articulated by Thompson (1967: 72–3), to quote a classic author in organization theories. Mass markets which are both homogeneous and stable breed a concentration of functions and more rigid bureaucracy.

The opposite situation exists in a heterogeneous and shifting market. A company may be seen to match such a market context with its own technical and organizational infrastructure in two main different ways:

> it may scrap old equipment and units and build up or add new ones, dedicated to new product lines within a more diversified and innovated product range;

it may also expand or diversify its product range by ma.. technical and organizational apparatus more flexible, withou. increasing the specialization and dedication of production lines or plants with regard to specific segments of the product range.

The two ways differ in their implications for the technical and organizational infrastructure. The first strategy represents a classical diversification and innovation by way of new, separate and fairly self-contained plants. Under the second strategy, the flexibility of one or more integrated plants is increased so that the dedication or product specificity of equipment and organizational units is reduced. The first strategy is well known, but less pertinent to the emergence of more differentiated product markets. It is more apt to cope with newly emerging homogenous markets. We therefore concentrate on the second strategy which appears to match more closely differentiating, heterogeneous markets.

The main link between properties of the product market and the organization of work and skills appears to be *batch size* – of products, components and parts – and strategies to translate product demand and variety into batches of components and parts. Experience from studies on new technology shows that when large batches are produced, the organizational and skills solution found is characterized by higher and more rigid segmentation of steps and jobs in the workflow, as well as by a more polarized distribution of skills; in the case of smaller batches, we find less segmented and more overlapping workflow organization and a less polarized distribution of skills (Sorge *et al.*, 1983). This effect extends through more or less the whole organization, not only to boundary-spanning components such as marketing, or research and development. The latter, more restricted and concentrated effect, appears to be correlated with the classical diversification mode referred to above which is less relevant here.

The effect we are dealing with is not so new, and it brings back the distinction between production continuity and workflow integration referred to earlier. Workflow integration can be seen as a concern to achieve high output in complex and capital-intensive systems. It may thus be a permanent feature of all technical change that produces complex and capital-intensive production systems. But production continuity is another thing: where batches are smaller, there is a constant need to retool, reset, replan, reprogramme, redesign and adapt to fluctuations of inputs due to discrepancies between planned or standard material quality, workpiece measures, design details and scheduling dates, on the one hand, and actual or unforeseeable values, specifications and dates on the other. Any reduction of the size of batches in a complex organization restricts the usefulness of constraining standards and central plans. This makes for workflow discontinuity, and thus gives rise to a need for developing and involving human competence, as both the socio-technical literature and industrial sociology in another tradition have shown (Lappe, 1986; *Human Systems Management*, 1986).

By relating organization and skills to product market properties, we do not intend to replace a technical with a market determinism. A company may look for a market niche to match its organizational strategy, or it may serve the market in a way that tallies with what it perceives to be its organizational and skill strengths. It can be shown that, for instance, the economic and employment success of the West German car industry in the 1970s and 1980s, when compared to the car industries of other countries, was strongly related to the pursuit of more qualitatively differentiated and quality-conscious markets. Success in such markets was conditional on the production of craft skills and their utilization for organizational flexibility. For this German firms, with the comparatively high skill level of their workforces, were particularly well placed. Moreover and at the same time, they were constrained by their works councils to increase their vocational training efforts in response to youth unemployment; to improve the quality of working life through task enlargement and task enrichment; to avoid redundancies through retraining and redeployment in a co-determined internal labour market, and so on. Given these pressures, it has been argued that, like the Swedish car manufacturers, they had little choice but to opt for an alternative to traditional price-competitive mass production (Streeck, 1986).

Long-range changes in organizational and qualificational strategies, and the resulting shift in the major explanatory approaches in industrial sociology and organization theories, cannot be due to a sudden, simultaneous increase in companies' awareness of socio-technical choice, or the potential in smaller and more differentiated markets. To assess the roots, the extent and the duration of factors which load socio-technical choice one way or the other, it is not enough to say that there is a choice, or to specify which contexts create which alternatives. We need a concept of how a society or economy is populated by technical and organizational types, and how this population changes over time. This has recently been discussed in the organizational ecology literature, by authors such as McKelvey (1982) or Trist (1981). We cannot comment on it adequately here, and we argue only that organization and skills are presently very much influenced by rationalization strategies aiming at a reduction of work-in-progress, stocks and throughput times, which, in the face of less homogeneous and less mass-type markets, reduces batch sizes of components and parts and thus feeds into organization and skilling policies.

Piore and Sabel (1984) have suggested that the revival of craft skills is linked to a transition from mass markets to more differentiated, quality-conscious and shifting markets. This development is not radically new, nor is it necessarily here to stay. We should guard against assuming this trend to be on-going, given the demise of past ideas about a continuing trend towards mass markets and the degradation of skills. We may be dealing with a cyclical phenomenon whose nature and causes are not known and hence require research. But the revival of interest in Kondratieff or innovation cycles has not been entirely helpful either, being confounded by a number of methodological errors and difficulties (Maddison, 1982). At the moment, one may only speculate (Firebaugh, 1983; Sorge, 1985; ch. 7) on the basis of limited empirical evidence and vague theoretical conjectures. We know that there are such cycles, but we know very little about their causes. However, despite their uniqueness and different duration, their phases seem to be governed by increasing returns to scale or scope, respectively.

To summarize the argument so far: the renewed emergence of shifting markets has put a premium on an organizational and skills formation strategy that often appeared to go against the tide in the post-war boom period. This has revived interest in Adam Smith's theory of the size of the market. Large homogeneous markets can be thought of as the corollary of a polarizing regime, whereas smaller and more heterogeneous markets are functionally related to a more organic socio-technical approach (Sorge, 1985; Warner, 1985). However, we do not suggest that organizations respond elastically and in the same way to a new or re-emerging set of conditions in goods markets. This is partly a matter of the extent to which a company already has some affinity with the appropriate strategy; or is turned round by the leading actors; or finds itself in a national-institutional context that has already made the respective strategy appear more opportune.

In our earlier research, we have indicated national differences that bear on the capacity to respond to new market conditions of differentiation and ability to change. The institutional context of Germany has favoured the development and quantitative proliferation of craft skills, less polarized distributions of skills, less job demarcation, and greater overlap between laterally and hierarchically differentiated work roles (Maurice *et al.*, 1980). Further below, we will demonstrate that the division of labour is closely related to the structure and processes of industrial relations. This will yield a concept of an interaction between organization, skilling and industrial relations which is not deterministic but rather emphasizes common logic or elective affinity that may be the result of reciprocal causation and interacting strategic choices.

Technical Change and Industrial Strategy

Close analysis of the relationship between technical change and the organization of work and skills draws attention to the impact of product markets and the formation of product strategies. But just as technology does not determine work organization, markets and products do not determine technology. New products tend to be accompanied by process innovations, and advances in process technology may enable firms to change to new products. But it may also be possible to produce a given product with different technologies, just as a given technology may be capable of producing a range of different products. In principle, then, there is always room for strategic choices between products and markets even without technical restructuring, and firms may also have a choice between different trajectories of technical development while continuing to produce for the same market. In this sense, technology and product are as loosely coupled, as are technology and work organization.

Technology does not determine industrial strategy – or, to use a term suggested by Willman (1985), manufacturing policy – but rather offers options from which management, trade unions and industrial policy-makers can select. This applies in particular to microelectronic circuitry which can be put to essentially two different uses, corresponding to different product strategies. On the one hand, it can be used for rationalization within traditional mass production, i.e. to save labour and to reinforce and extend the separation of conception and execution. Used in this way, microelectronics enables producers to supply standardized products at lower prices, thus increasing or restoring their competitiveness in mass markets. In addition, it seems that microelectronic technology has changed the economies of scale in mass production, in that it enables manufacturers to reach break-even point with much smaller batches than in the past. This has helped smaller firms survive in industries such as automobiles which, according to widely accepted predictions, should by now consist of no more than three or four mega-producers (Altshuler *et al.*, 1984: 181ff.). For the same reason, new technology facilitates reorganization strategies that rely heavily on capacity cuts, down-scaling of operations, and the shedding of labour.

Yet at the same time microelectronics can also be used to rid manufacturing industries of the principal limitation of traditional mass production: the high rigidity of production equipment. Microelectronic equipment can be designed to be far less dedicated to given products than equipment automated on the basis of conventional electronics, electromechanics, mechanics, hydraulics, etc. Since machinery equipped with microelectronic controls is easily retooled, it can be used for the highly diversified production of individualized goods, at costs rising less than with previous purpose-specific technology. In this sense, microelectronics, in addition to changing the economies of scale, makes

it possible to reap increasing returns to scope (Sorge, 1'
the introduction of high product variety in large-scale
processes. Moreover, microelectronic production equipme.
used to improve product quality. The resulting type of custoი.
diversified high-quality products responds to non-mass markets ᵢ.
which competition is not only over the price of basically homogeneous
goods but over product quality and the degree to which products meet
the special needs of individual customers.

A more systematic analysis of the alternative manufacturing policies
from which new technology enables firms to choose, would probably
emphasize three variables: the degree to which products are standardized; the type of competition to which they are exposed; and the
volume in which they are produced. The first two factors seem to be
closely related, in that standardized products are generally sold in
price-competitive markets, whereas customized products tend to be
quality-competitive. This suggests a distinction between *standardized
price-competitive* and *customized quality-competitive* production, on the
one hand (Cox and Kriegbaum, 1980), and *low* and *high volume*
production, on the other. Production volume is, of course, a function
of production unit size and factor productivities. Since we do not
envisage a drop in labour productivity, but only slower increases, and
since size of production units and standardization of products cannot
be conflated, it is important to distinguish the two dimensions, of
volume and of standardization of production. Small batches may be
beautiful in small or large units, with small or large volume alike. The
same applies if beautiful is replaced by ugly.

Crossing the two dimensions, one arrives at four alternative product,
or manufacturing, strategies (see figure 1.1) whose relationships with
technology seem to be central for understanding present processes of
industrial restructuring. Of the four possible types of manufacturing
strategies defined by the two dimensions of our contingency table, two
– the low-volume production of customized quality-competitive goods,
and the high-volume production of standardized price-competitive
goods – look familiar. Indeed, with some simplification one could say
that before the advent of microelectronic technology, theirs would
have been the only cells in the matrix that would have been filled,
apart from the small suppliers of specialized components to mass
production that might be seen as inhabiting cell 1. This comparatively
simple picture has now become considerably more complicated. For
example, it has already been mentioned that new technology may have
lowered the break-even point of mass production, thus eroding
somewhat the boundary between cells 3 and 1. On the other hand, a
small component producer (cell 1) dependent on a large assembler
may under the new technical conditions find it attractive to move into

	Standardized Price-Competitive Products	Customized Quality-Competitive Products
Low volume	Specialized component production 1	Craft production 2
High volume	3 Mass production ('Fordism')	4 Diversified quality production

FIGURE 1.1
A Simple Classification of Product Strategies

craft production (cell 2) by developing and differentiating his product range, so as to become less dependent on price fluctuation and monopsonistic demand. Moreover, by lowering the costs of customized high-quality production, new technology has not only opened new avenues of expansion for previously dependent suppliers but has also made it possible for small specialist producers to remain, or again become, economically viable (cell 2), which may put a brake on, or reverse, the concentration of manufacturing industries.

The most important impact, however, of new technology on manufacturing strategy seems to be that it has made it possible to fill another cell in our matrix by creating a new option of high-volume production of customized quality-competitive goods (cell 4). In many sectors of manufacturing industry, microelectronic circuitry has progressively eroded, in the course of the past decade, the traditional distinction between mass and specialist production. The extreme flexibility of microelectronic equipment, and the ease and speed with which it can be reprogrammed, have enabled firms to introduce a hitherto unknown degree of variety, as well as quality, in large batch production. The result is a restructuring of mass production in the mould of customized production, with central features of the latter being blended into the former and with small batch production of highly specific goods becoming enveloped in large batch production of basic components or models. We prefer to call this new pattern diversified quality production since the more general term, flexible specialization (Piore and Sabel, 1984), has become too closely associated with the notion of small, independent craft firms (cell 2)[1]. It can be approached via two different trajectories of industrial restructuring: by craft producers extending their production volume without having to sacrifice their high quality standards and customized product design (moving, so to speak, from cell 2 into cell 4) or by mass producers moving upmarket by upgrading product design and quality, and by increasing product variety in an attempt to escape from the pressures of price competition and from a market segment which seems to be becoming smaller (starting, as it were, from cell 3). Industrial restructuring towards diversified quality production is now generally regarded as a highly promising strategy for old industrial, high wage economies striving to remain competitive in more volatile and crowded world markets, while at the same time trying to protect their employment in manufacturing.

Firms in cells 2 and 4 operate and thrive where the premise of increasing returns to scope applies, i.e. where it pays to differentiate and upgrade the product range. Firms in cells 1 and 3 operate and thrive where increasing returns to scale are seen to exist, i.e. where it pays to specialize and standardize production. Successful operation in cells 2 or 4 may lead to a growth in market share which may be linked to a growth of volume. This might move the firm away from cells 2 or 4, but only if a growing market share is not linked to growing product market differentiation. Therefore, while it is conceivable that success in one cell may lead to a movement to the boundary of that cell or beyond, neither the existence nor the direction of such a movement can be taken as given. They depend on how an economy's or society's ecology of product markets evolves, and how this is interpreted by crucial actors. Moreover, even if shifts of emphasis between cells occur, there is not necessarily a zero-sum game between firms with different manufacturing strategies. An increasing population of cells 2 and 4, for instance, may require an increasing population of cell 3 by manufacturers of standardized and mass-produced multi-purpose electronic devices, such as memories and microprocessors which are required for flexible control equipment.

New technology, since it may improve firms' survival chances in any of our four types of production, does not as such determine manufacturing policy. If at all, it seems to extend rather than narrow down the range of available choices. Which package of products and processes a firm selects depends above all on the opportunities it perceives in product markets. But since production technology and the way in which it is utilized is also related to the organization of work, the choice of new manufacturing policies is in addition conditioned by, and may require restructuring of, the socio-technical production system. The latter, in turn, relates back to the labour market and the supply of skills that is on offer.

The Role of Industrial Relations

Up to now, we have seen technical change interacting closely with both the organization of work and skills and firms' strategic choices of products and markets. In particular, we have seen that the way in which technology is integrated with work organization in socio-technical systems is conditioned by the markets to which industrial organizations (choose to) respond. We now proceed to discuss how industrial relations fit into the picture.

The principal bread-and-butter themes of industrial relations have always been wages and conditions, with macroeconomic demand management and control of inflation supplementing the discipline's original microperspective in the corporatist 1960s and 1970s. According to Dunlop's seminal definition of the 'industrial relations system' (1958), technology is not a *subject* of joint regulation but one of its three main *environments*, the other two being the market and the distribution of power in the larger society. Just as these, technology, or technical change, affects the operation of industrial relations systems from the outside. The output, as it were, of industrial relations systems consists of a 'web of rules' that governs the workplace and the work community, i.e. the organization of work. To the extent that technical change requires changes in work organization, it may bring pressure to bear on rule-making bodies to negotiate, impose or accept normative adjustments. It is also conceivable in the model that rigid or inert rule systems, perhaps defended by one or another interested party, stand in the way of the introduction or full utilization of particular new technologies. In any case, however, technology as such remains an exogenous factor which in itself is outside the scope of functionalist-pluralist rule-making.

It is not a long step from here to the notion that technology just as, for example, product strategy, is a managerial prerogative, and that the role of industrial relations with respect to technical change is properly restricted to regulating its consequences in the workplace, but not its substance or direction. This view, in turn, is closely linked to an idea of technical development as an essentially unilinear and unidimensional process which is either interest-neutral (the pluralist industrialism tradition – Kerr *et al.*, 1960) or closely identified with the interests of management (the radical position as held by Braverman, 1974, and Fox, 1974). The consequence is the same: *there is in principle nothing to negotiate.*

It is only consequent on this background that in mainstream industrial relations research and theory, the themes of technology and technical change appear almost exclusively in terms of resistance to change or acceptance of innovation (Willman, 1985). For the practical art of conducting good industrial relations, the task was essentially to design pluralist procedural rules that facilitated the consensual, frictionless adaptation of substantive rules to whatever the exogenous process of technical development required. Among the main preoccupations of the discipline in this respect has been the comparison of different industrial relations systems in terms of their openness to technical change. Among the variables that were used to distinguish industrial relations systems for this purpose are:

the extent to which workers are represented by craft unions organizing only skilled workers of specific occupations, as opposed to company or industrial unions representing workers of all occupations and skill levels in a given firm and industry;

the degree to which negotiations are centralized above the shopfloor level, i.e. are conducted for entire companies, industries, or nations;

the relative importance of job control as compared to industrial democracy, i.e. the extent to which rights of workers to participate in managerial decisions are located on the shopfloor and relate to the design of individual jobs, rather than being institutionalized at the enterprise level and referring to the firm's overall commercial strategy;

the role of the external as distinct from the internal union in collective bargaining, in particular the degree to which collective bargaining is controlled by full-time union officials outside the individual firm;

the degree to which industrial relations are conducted under mutually recognized (formal or informal) rules of the game which separate clearly the domains of joint regulation from those of managerial prerogative;

the extent to which substantive matters such as working conditions, employment and social security are regulated by legislation as distinct from industrial agreement, and the extent to which trade unions as a consequence (have to) rely on political as opposed to industrial action.

Underlying these distinctions, and apparently confirming their validity, was the following observation: firms, industries or countries with industrial or company unions, relatively centralized negotiations, industrial democracy, well-formalized demarcation of spheres of influence for management and trade unions, etc. experienced less resistance to, and more consensus on, technical change than was found in the absence of these conditions (see, for example, Jacobs *et al.*, 1978; Hotz-Hart, 1987; Maitland, 1983). Although in some respects the evidence was not unambiguous, in the United Kingdom in particular it provided the basis for a liberal–pluralist programme of industrial relations reform which culminated in the Donovan Report of 1968, calling for more orderly industrial relations as a precondition for improvements in economic performance through technological and organizational modernization.

The idea of institutional reform was closely linked, in the UK and elsewhere, to the identification during the 1960s and 1970s of various models of industrial relations systems that were believed to be ideally suited, among other things, to accommodating technical change: the United States with its high degree of legal formalization; Sweden with its encompassing industrial unions; West Germany with its peculiar institution of co-determination; and, lately, Japan. However, all attempts to transfer model institutions of industrial relations from one country to another have failed. Radical opponents of the liberal Oxford School in Britain have explained the persistent disorder, in the face of repeated attempts at reform, with reference to the fundamental inequality between capital and labour which renders the achievement of distributive consensus impossible. As its key witness, the critique of industrial relations institutionalism adopted the early Durkheim, the author of the famous verdict in *The Division of Labour* that 'there cannot be rich and poor at birth without there being unjust contracts' (Durkheim, 1964: 384). This, it was suggested, was why capitalist societies are beset by inherent anomic tendencies which are the real causes of, among other things, the rising rate of inflation (Fox, 1974: 322; Goldthorpe, 1974). Persuasive as this argument may appear, however, it does not provide the conceptual equipment to account for cross-national differences in either industrial relations conflict (Maitland, 1983) or differing inflation rates.

On the other hand, in its most advanced version (Fox, 1974) the radical critique of pluralist institutionalism raises the theme of technology in a way which is highly instructive. Industrial anomy, according to Fox, is the result of a proliferation of 'low trust' which starts in the Taylorist organization of work. Fox assumes, and never doubts, that to achieve efficiency managements have to organize labour in such a way that tasks become ever more standardized and subdivided (Fox, 1974: 59, 96, *passim*). In a brilliant analysis that mobilizes the power of a long sociological tradition, Fox proceeds to argue that the persistent efforts of the high-discretion fraternity of managers to curtail the discretion vested in the work roles of their organizations' lower participants, are bound to erode the capital of trust and goodwill that is a necessary precondition for contractual relationships to function (Fox, 1974: 87f). A Taylorist organization of work – as required by the economic imperatives of mass production – thus generates its own conflicts and crises in that it gives rise to calculative and irresponsible attitudes on the part of workers (Fox, 1974: 64, *passim*). It also induces workers to try to defend themselves collectively by imposing in turn similar limitations on managerial discretion through formalized, specific rules and procedures (Fox, 1974: 113). It is here that, among other things, organizational rigidities and restrictive practices against technical change arise as rational responses by workers to their subordinate position in the organization of work.

Fox's criticism of the Oxford School appears convincing enough where it points to the futility of procedural–institutional reform as a means of curing problems that are deeply rooted in work organization and technology. At the same time, if the assumption of the coincidence of economic rationality with mass production and Taylorism were relaxed – as we have strongly suggested it should be – the movement towards industrial anomy would look somewhat less inevitable, and it might also appear possible to explain different *degrees* of anomy and conflict in industrial relations by differences in socio-technical systems and the generation and utilization of skills. Fox, however, does not follow this path and cannot do so given his unquestioned acceptance of the mass production view of technology and efficiency. To him, cross-cultural differences in work organization are basically due to time lags in the modernization process, and indeed in a special section on Japan he predicts that the more organic Japanese organization of work will shortly fall victim to economic pressures for Westernization (Fox, 1974: 131-5). It is hardly necessary to say that today the pressure goes exactly in the opposite direction. While Fox is undoubtedly right where he refutes the idea that industrial inefficiency could be remedied through better procedures and institutions of industrial relations, he ignores the possibility that the crisis of both industrial relations and industrial efficiency in Britain in the 1970s may have been a result of particularly heavy reliance on Taylorist modes of (low-skill) work organization, combined with and in response to historically privileged access and prolonged exposure to mass (world) markets.

Turning to technical change in particular, the liberal reform tradition operated on the premise that resistance to change, and insistence on inefficient restrictive practices, is caused by informal, fragmented, decentralized, disorderly industrial relations privileging conservative short-term over enlightened long-term interests of workers. In opposition to this view, it has been argued that the interests pursued by such unions are quite rational, given the organization of work and skills in which they have to operate. While we tend to support the second position, we are not discarding altogether the first, Olsonian one. Indeed, we believe that both may be valid at the same time. While a fragmented, sectional, job control-based industrial relations system may give rise to rigidities in work organization that impede technical change, the opposite effect may also exist: industrial relations rigidities, rather than being the source of adjustment problems of work organizations to new technology, may simply reflect them.

For example, a high degree of separation of conception and execution inside the organization of work is likely to give rise to contests over managerial prerogative which would not emerge where the differentiation of functions is less pronounced. This applies in particular when, in the course of introducing new technology, the existing division of labour needs to be reorganized. While in a more organic work organization, adjustments to technical change can often be made within individual work roles, under high functional differentiation between management and the shopfloor, adjustment problems tend to be transferred into the industrial relations system where they are typically transformed into matters of pay and conditions. In this sense, systems of work organization with much management tend to have more industrial relations, – seen as a specialized area outside production management – than organic systems with functionally diffuse work roles.[2]

The crucial factor, as we have pointed out in earlier work (Sorge and Warner, 1980; Streeck *et al.*, 1981), is the way in which skills are defined, generated, and organized into occupational career trajectories. Industrial unionism and the accompanying willingness of unions to accept technical change are conditional upon some degree of mobility between occupations; this, in turn, presupposes broad general skills that provide a basis for further training and reversible specialization. Craft unionism defends rigid inter-occupational boundaries, which correspond to narrow skills that are difficult to adjust. Where the latter type of skill prevails, workers will tend to define qualification as an entitlement to perform certain, fixed and specific tasks. Firms, on the other hand, will depend for adjustment to technical change on a managerial prerogative to hire and fire workers as the only way to change the skill composition of their workforce. In response to the high external flexibility that firms are forced to defend, workers will insist on strong internal rigidities protecting the tradeability of their skills on the external labour market. External flexibility and internal rigidity, in their affinity to narrow skills, a strong external labour market and a high division of labour, are compatible with economic performance as long as there are enough skilled workers available on the external labour market, and employed workers lack the power or the incentive to constrain firms' ability to hire and fire.

The situation is quite different where skills are broad enough to permit extensive internal retraining and redeployment. In this case, workers will find it less necessary to engage in sectional defence of job territories. Since the external transferability of their broader skills is not damaged by redeployment and retraining inside their present organization of work, they can afford to accept a flexible allocation of work tasks. This, in turn, makes it possible for management to make

concessions on employment security, and for workers to 〈
interests in terms of imposing external instead of internal r
management. A flexible internal labour market in a less 1
differentiated organization of work also invites trade union participation
in the management of redeployment, that is some form of shared
responsibility under industrial democracy or co-determination for an
efficient functioning of the internal labour market (Streeck, 1986).
Moreover, broadly skilled workers, being in a position to give up job
control in favour of a combination of external rigidity and internal
flexibility, are likely to find their interests most adequately represented
by unions which negotiate general rules – susceptible to and, in fact,
in need of formalization – rather than locally idiosyncratic, informal
job territories. This in turn is conducive to trade union reliance on
political action, and to incorporation of trade unions in public policy
and public responsibility. Where the division of labour and authority
is less developed, defence of the interests of workers can be less
localized, more easily subject to formal and general rules, and interest
representation can be removed from the shopfloor to higher levels of
collective action. It is against this background that the paradoxical
coincidence of extensive legal regulation with high shopfloor flexibility
in Germany, and voluntary industrial relations with extreme rigidities
in the organization of work in the United Kingdom (at least until the
early 1980s), might be explained.

Again, there is a connection with product markets and product
strategies. To Anglo-Saxon observers in particular, one of the
extraordinary features of German trade unions is their untiring support
for, and active political commitment to, an expansion and upgrading
of vocational training. This is in spite of the fact that from a pure
labour market perspective, any increase in the supply of skilled labour
is bound to depress its price, which is precisely why craft unions in
the United States and the United Kingdom have always tried to limit
the number of apprentices in their respective trades. The reason why
German trade unions apparently act against their own interests is their
strong awareness of the exposure of German industry to a highly
competitive world product market. As latecomers to industrialization
who had to find opportunities for expansion outside the protected
markets of established mass producers, German firms have from early
on specialized in diversified quality products, with high service intensity
and customized design. It was this characteristic market strategy that
made it possible for employment in German manufacturing to remain
more stable than in any other old industrial country, in spite of trade
unions having learnt over time to expect and extract high real wages.
Today, German unions in exposed sectors are conscious of the fact
that high-wage manufacturing industries in old industrial countries can
survive only with a flexible quality production system based on a large

supply of broad, adjustable skills. This is why they support a training
system which, precisely because it tends to produce an oversupply of
skills, facilitates an organization of work that can relatively easily cope
with technical change (Streeck, 1986). And it is this configuration in
the organization of work and skills which both sustains and is sustained
by an industrial relations system that is non-sectional, encompassing,
formalized, legally regulated, and centralized.[3].

National patterns of skill formation represent an important element
of the societal effect by which Maurice *et al.* (1980) have explained
cross-cultural differences in the division of labour and the organization
of work. To the extent that such differences are, as we hope to have
shown, related to different patterns of industrial relations, they appear
to offer a better explanation than that suggested by the radical
Durkheimians for the lack of success of liberal-reformist projects to
transplant institutions of industrial relations from one country to
another. Liberal–pluralist institutionalism seems to be frustrated not
so much by a general trend towards low trust and anomy in capitalist
society, as by the complex interdependence of industrial relations
institutions with a wide range of economic, social and cultural factors.
Together these seem to form distinctive configurations of interdependent
elements – historical individuals, in Max Weber's terminology –
that may operate as vicious or virtuous circles depending on the
circumstances, and that seem to be highly resistant to purposive,
monocausal intervention due to their inextricable internal complexity.

Conclusion: Technical Change, Industrial Relations and Trade Unions

Technology and technical change interact with both a firm's product
strategy and its work organization (figure 1.2). By interaction we mean
a loosely coupled relationship of mutual dependence in which directions
and mechanisms of causation may vary under different internal
and external conditions (Weick, 1979). Moreover, the density and
determinacy of the interrelation beteen product, process and work
organization are in part a matter of strategic organizational choice, as
each of the three can be designed to be more or lesss autonomous of
the others. Both product strategy and work organization, in turn, are
related to crucial organizational environments – the former to the
product and the latter to the labour market, including the qualificational
structure of the labour supply and the societal system of skill formation.
Here again the relationship seems to be one of mutual interaction
rather than unilateral dependence (for a more detailed discussion of
the model, see Streeck 1985).

Management and trade unions – including institutions of workplace

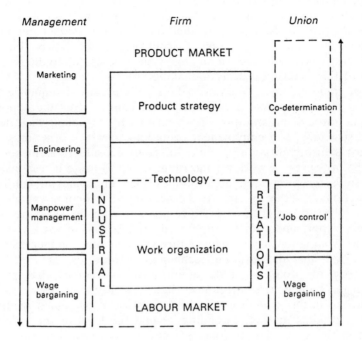

FIGURE 1.2
Technology and Industrial Relations in Context

representation such as works councils and shop stewards – relate to
the two markets and the three areas of strategic organizational choice
in significantly different ways. In a simplified and ideal–typical model
– which is to serve as an analytical baseline for a more complex picture
of the politics of the relationship between product and labour markets
– the interests of managements and unions have in common that they
are directed at reducing the complex set of interactive relationships
(summarized in figure 1.2) to a sequential order of unilateral dependence
constituting a consistent hierarchy of causal determination and decision-
making criteria. Interests differ, however, in that the two sides'
preferred decision sequences run in opposite directions. For the
management of a firm operating in a competitive market environment,
product strategies have to follow the signals of the market; technology
has to follow both product design and competitive (market) pressures
for low production costs; and work organization has to be fitted to the
adopted technology so that the resulting products can optimally exploit
the opportunity structure of the market. Finally, how much and what

kind of labour is hired, and at what price, is determined by the requirements of an optimal organization of work, and it is the task of collective bargaining to ensure that the labour market supplies exactly what the firm needs. Marketing, therefore, governs product and process engineering, and these together control the manpower function, including collective wage bargaining, which is relegated to the receiving end of the managerial decision-making process.

For trade unions, by contrast, the ideally preferred sequence takes off at the labour market. By acquiring control over the supply of labour through organization, trade unions try to define wage levels, skill structures and employment conditions to which firms must then adjust their organization of work. Job control inside firms further adds to the pressures on work organization, trying to make it responsive to demands for promotion opportunities, skill enlargement, control of discretionary authority etc. At the next level, technology would then have to be so designed as to be compatible with the structure of the labour supply and with the preferred organization of work (which is in fact what unions try to achieve through technology agreements). To the extent that the resulting technology is dedicated to a particular category of products, trade union interests require that product strategies be matched to technical choices, rather than vice versa. The final step in this sequence would be a system of bargained regulation of the product market, extending beyond the level of the individual firm in the same way as collective bargaining extends to the labour market. Such systems, however, do not normally exist, and it is here that the apparent symmetry of the positions of management and trade unions comes to an end.

Trade unions, unlike management, are formally involved in a firm's decision-making only as far as it relates to the firm's external and internal labour markets, and the linking of the two through hiring, redundancy and training policies. As a result, important parameters affecting the choice and design of new technology are outside their institutionalized sphere of influence. Lacking a collectively bargained product market, trade unions are limited to creating institutional rigidities at the lower end of the managerial decision-making sequence that constrain management's decisions on pay, work organization and, in part, technology. While superficially these rigidities correspond to those of the product market represented by management and constraining trade unions, the primacy of the product market over the labour market, far from being merely the preference of one of two rivalling parties, is strongly institutionalized in the structure of a competitive market economy. It is also, and correspondingly institutionalized in the limited domain of the functionalist–pluralist subsystem of industrial relations, as represented by the dotted line in figure 1.2. Product strategy, which affects technology as much as does the organization of

work, is *not* included in the area of joint regulation since
to be exclusively governed by the product market. An
asymmetry is, of course, that inside the industrial r
trade unions, unlike management at the firm's p.
boundary, have to share power with management under joint regulation.

The power of trade unions to create rigidities in the labour market
is conditional on the structure and level of demand in the product
market, whereas the power management derives from representing the
imperatives of the product market is independent of the labour supply.
If product demand is high, demand for labour is also high, and trade
unions can successfully press for regulations that limit managerial
discretion on employment and working conditions. But if products no
longer fit the market, trade union power is bound to wane, and
management may be presented with the opportunity to roll back labour
market rigidities in a way which in a market economy is not normally
available to trade unions in relation to product market rigidities. Due
to the hierarchical superiority of product over labour markets,
maladjustment of firms or industries to product demand constitutes a
source of power for management and enables it to reaffirm the
hierarchical sequence of control from product strategy to technology,
work organization and finally, to labour demand. Maladjustment to
the supply of labour, on the other hand, does not bestow comparable
power on trade unions, and it is this fundamental asymmetry that gives
rise to the structural inferiority of trade unions in a capitalist economy.

But this is not the whole story. Managerial product strategies –
manufacturing policies – respond to the market but are not determined
by it. Market signals are never decisive, partly because active marketing
may be able to change them, and partly because even the best
marketing department cannot safely predict where future profits will
be found. Moreover, there is always a variety of markets from which
firms can choose, the most basic distinction being that between mass
and specialized, or price- and quality-competitive, markets. Firms also
have a limited but nevertheless significant degree of choice with regard
to their own performance standards, with firms differing among other
things along organizational, sectoral and national lines in terms of the
time they are willing to wait for investment to become profitable.
Finally and most importantly, economic and technical change, as we
have argued above, has enlarged the matrix of strategic choice for
firms in manufacturing and has simultaneously made the coupling
between central areas of choice, such as product strategy, technology
and work organization, less determinate and richer of functional
alternatives and equivalents.

Uncertainty and indeterminacy create space for alternatives in
commercial strategy which make it possible, and inevitable, for
managements to take into consideration factors other than the market.

Previous investment in a specific manufacturing organization and technology may commit managements to a particular product strategy. Trade unions and industrial relations may also be important in that they affect the available skill mix and the relative prices of labour. Since the market cannot fully determine strategic production decisions, the rigidities created by trade unions at the labour market end of the managerial decision-making hierarchy operate as a set of constraints and opportunities that influence strategic managerial decision-making – if only inadvertently. If a product strategy that has been adopted in an attempt to accommodate existing organizational structures and labour market conditions yields satisfactory results by the firm's culturally and politically defined performance standards, and is expected to remain successful, there is no need for management to challenge industrial relations rigidities even if they prevent it from pursuing alternative strategies. It is towards these multi-causal and interdependent relationships that trade union policy might turn for re-orientation in an era of extended and renewed strategic choice in both manufacturing policies and industrial relations (Kochan *et al.*, 1984).

Many of the problems that trade unions face in the 1980s result from attempts of managements to erode the boundaries between firms' commercial experience and the institutional subsystem of industrial relations whose relative autonomy from fluctuations in the economic situation of enterprises and industries was established in the post-war period (Strauss, 1984). Underlying such efforts are fundamental changes in the economic environment which require firms to react more directly to more volatile market conditions. One way of becoming more flexible is by integrating industrial relations and manpower policy into comprehensive commercial strategies, tying the functionally differentiated areas of marketing, product design, process engineering, work organization and skill formation more closely together. Tendencies of this kind are almost universal in the present restructuring process, and it seems that their wide presence is related above all to the potential of new technology for more flexible manufacturing.

For trade unions, the traditional separation of industrial relations from marketing, finance, and product and process engineering once offered a degree of shelter and protection for independent action in the defence of their members' interests. However, now that it has become increasingly unlikely that the subsystem boundaries of the 1950s and 1960s can be defended or re-established, independence may increasingly come to mean irrelevance in the face of a much more tightly co-ordinated managerial decision-making process. The only alternative for trade unions may be to develop an integrated strategy of their own, by extending their strategic concerns upwards beyond the traditional subsystem limits of industrial relations. In trying to do this, they can be supported by more careful empirical examination, not least on a comparative, cross-cultural basis, of the impact, hitherto largely unintended and latent, of trade union presence and action on the conception of commercial strategies – which might enable them to put the respective causal links and connections to strategic use.

Workers' control of industry, as a form of industrial democracy that extends beyond negotiations of pay and conditions to include the politics of production, was a radical project in the 1960s and 1970s, allied with a perspective beyond capitalism. It was as much a reaction against mainstream industrial relations, with its functionalist perspective and its functionally differentiated area of concerns and mode of operation, as it tried to formulate an alternative to the capitalist mode of production. Today, it is generally viewed as a marginal and rather unsuccessful movement which has more or less withered away.

But this is not what our analysis suggests. Giving up the comprehensive perspective of the radicals just because of the turning of the political tide in the 1980s may be premature. This is because there has also been a turning of the tide in product markets that puts a premium on a type of industrial relations in which trade unions are both constrained and offered the opportunity to bear a much more conscious and explicit regard for product, marketing and technical strategies. In many ways, this development seems to vindicate yesterday's labourist, syndicalist and socialist doctrines that have unceasingly insisted on a strong link between both the politics of labour markets and of production, both as a precondition of effective representation of worker interests and as a strategic lever for societal transformation.

At the same time, however, it also appears that the turning of the tide, both political and economic, has taken the radicalism out of what was once a project of fundamental political change. What may initially have been a revolutionary idea seems to have become, under new economic conditions, a necessary requirement of competitive economic performance in world markets. Unions moving in the direction of a comprehensive production strategy away from mass production are likely to help capitalist firms link successful marketing and production with peace at the workplace. Once more, Janus appears to rear his head, and to the radicals it may this time be a particularly ugly one: a programme of fundamental social change performing a latent function for social system integration – and possibly also for social value integration which, *pace* Lockwood (1964), may not after all be so independent of the former.

Notes

A version of this chapter was presented as a paper at a German–Japanese workshop, 'Coping with New Technology in Japan and the Federal Republic of Germany', in Berlin, August 1986. We are grateful to the participants and to a number of colleagues at the Wissenschaftszentrum Berlin for detailed and constructive criticism.

1. In fact, we are arguing that precisely because new technology enables large firms to avail themselves of the organizational means to render their own design and production more flexible, their chances of competing in the changed ecology of product market types may not at all suffer. There is good reason indeed to guard against an unbridled small is beautiful philosophy.

2. This may be one reason why, in a country like Germany, the very concept of industrial relations has failed to take root. Another case in point is the often observed low (undeveloped) differentiation of company unions from lower and middle management in the organic work organization of Japanese firms, where many of the functions of both management and trade unions are incorporated in the work group (Deutschmann, 1986).

3. It is obvious that our analysis draws heavily on the German case and on its difference to countries like the United Kingdom or the United States. But we believe that with appropriate modifications, the argument can easily be extended to other countries such as, for example, Japan or Sweden on the one hand and France and Italy on the other.

References

Altshuler, A., M. Anderson, D. Jones, D. Roos, and J. Womack. 1984. *The Future of the Automobile: The Report of MIT's International Automobile Program*. Cambridge, Mass.: MIT Press.

Batstone, E., A. Ferner and M. Terry, 1984. *Consent and Efficiency: Labour Relations and Management Strategy in the State Enterprise*. Oxford: Basil Blackwell.

Blau, P.M., and R.A. Schoenherr, 1971. *The Structure of Organizations*. New York/London: Basic Books.

Blauner, R. 1964. *Alienation and Freedom: The Factory Worker and his Industry*. Chicago: Rand-McNally.

Braun, E., and P. Senker, 1982. *New Technology and Employment*. London: Manpower Services Commission.

Braverman, H. 1974. *Labor and Monopoly Capital. The Degradation of Work in the Twentieth Century*. New York: Monthly Review Press.

Burns, T., and G.M. Stalker, 1961. *The Management of Innovation*. London: Tavistock.

Child, J., and R. Mansfield, 1972. 'Technology, Size and Organization Structure'. *Sociology*. Vol. 6, 370–93.

Cox, J.G., and H. Kriegbaum. 1980. *Growth, Innovation and Employment: - An Anglo–German Comparison*. London: Anglo–German Foundation for the Study of Industrial Society.

Deutschmann, C. 1986. 'Economic Restructuring and Company Unionism: The Japanese Model'. Wissenschaftszentrum Berlin: Discussion Paper IIM/ LMP 86–17.

Dunlop, J. 1958. *Industrial Relations Systems*. Carbondale and Edwardsville: Southern Illinois University Press.

Durkheim, E. 1964. *The Division of Labor in Society*. New York: Free Press.

Firebaugh, G. 1983. 'Scale Economy or Scale Entropy? Country Size and Rate of Economic Growth, 1950–1977'. *American Sociological Review*. Vol. 48, 257–69.

Fox, A. 1974. *Beyond Contract: Work, Power and Trust Relations*. London: Faber.

Friedmann, G. 1950. *Où va le travail humain?* Paris: Seuil.

Gensior, S. 1986. 'Mikroelektronik-Anwendung und ihre Bedeutung für die Qualifikation. Ein Literaturbericht.' Wissenschaftszentrum Berlin: Discussion Paper IIM/LMP 86–7.

Goldthorpe, J.H. 1974. 'Social Inequality and Social Integration in Modern Britain'. *Poverty, Inequality and Class Structure*. Ed. D. Wedderburn. Cambridge: Cambridge University Press.

Gunn, T.G. 1982. 'The Mechanization of Design and Manufacturing'. *Scientific American*. Vol. 247, 87–108.

Heydebrand, W.V. 1973. 'Autonomy, Complexity, and Non-bureaucratic Coordination in Professional Organizations'. *Comparative Organizations. The Results of Empirical Research*. Ed. W.V. Heydebrand. Englewood Cliffs, NJ: Prentice-Hall, 158–89.

Hotz-Hart, B. 1987. *Modernisierung von Unternehmen und Industrien bei unterschiedlichen industriellen Beziehungen*. Bern, Stuttgart: Haupt.

Human Systems Management. 1986. Special issue 'Human Resources in the Computerized Factory'. Ed. D. Gerwin, A. Sorge, and M. Warner. Vol. 6, no. 3.

Jacobs, E., S. Orwell, P. Paterson, and F. Weltz. 1978. *The Approach to Industrial Change in Britain and Germany*. London: Anglo–German Foundation for the Study of Industrial Society.

Kern, H., and M. Schumann. 1970. *Industriearbeit und Arbeitsbewusstsein*. Frankfurt a.M.: Europäische Verlagsanstalt.

Kern, H., and M. Schumann. 1984. *Das Ende der Arbeitsteilung? Rationalisierung in der industriellen Produktion*. München: C.H. Beck.

Kerr, C., J.T. Dunlop, F.H. Harbison, and C. Myers. 1960. *Industrialism and Industrial Man*. Cambridge, Mass.: Harvard University Press.

Kochan, Th. A., R.B. McKersie, and P. Cappelli. 1984. 'Strategic Choice and Industrial Relations Theory'. *Industrial Relations*. Vol. 23, 16–39.

Lappe, L. 1986. 'Technisch-arbeitsorganisatorischer Wandel und seine Auswirkungen auf die Beschäftigten in der Metallindustrie'. *Auswirkungen neuer Technologien*. Ed, E. Ulrich and J. Bogdahn. Ergebnisse eines IAB-Seminars. Nürnberg: Institut für Arbeitsmarkt- und Berufsforschung, Beitr AB 82.

Lockwood, D. 1964. 'Social Integration and System Integration'. *Explorations in Social Change*. Ed. G.K. Zollschan and W. Hirsch. London: Routledge & Kegan Paul, 244–57.

McKelvey, W. 1982. *Organizational Systematics*. Berkeley: University of California Press.

Maddison, A. 1982. *Phases of Capitalist Development*. Oxford: Oxford University Press.

Maitland, I. 1983. *The Causes of Industrial Disorder: A Comparison of a British and a German Factory*. London: Routledge & Kegan Paul.

Manpower Services Commission. 1985. *The Impact of New Technology on Skills in Manufacturing and Services. A Review of Recent Research*. London: Manpower Services Commission.

Maurice, M., A. Sorge, and M. Warner, 1980. 'Societal Differences in Organizing Manufacturing Units. A Comparison of France, West Germany and Great Britain'. *Organization Studies*. Vol. 1, 59–86.

Parker, S.R., R.K. Brown, J. Child, and M.A. Smith, 1972. *The Sociology of Industry*. London: Allen & Unwin.

Piore, M. and C. Sabel, 1984. *The Second Industrial Divide*. New York: Basic Books.

Sorge, A. 1979. 'Technical Change, Manufacturing Organization and Labour Markets. Effects and Options of Microelectronics'. Wissenschaftszentrum Berlin: Discussion Paper IIM/LMP 79–15.

Sorge, A. 1985. *Informationstechnik und Arbeit im sozialen Prozess. Arbeitsorganisation, Qualifikation und Produktivkraftentwicklung*. Frankfurt a.M.: Campus.

Sorge, A., and M. Warner, 1980. 'Manpower Training, Manufacturing Organization and Workplace Relations in Great Britain and West Germany'. *British Journal of Industrial Relations*. Vol. 18, 318–33.

Sorge, A., G. Hartmann, M. Warner, and I. Nicholas, 1983. *Microelectronics and Manpower in Manufacturing*. Aldershot: Gower.

Stinchcombe, A. 1959. 'Bureaucratic and Craft Administration of Production: A Comparative Study'. *Administrative Science Quarterly*. Vol. 3, 168–87.

Strauss, G. 1984. 'Industrial Relations: Times of Change'. *Industrial Relations*. Vol. 23, 1–15.

Streeck, W. 1985. 'Introduction: Industrial Relations, Technical Change and Economic Restructuring. Industrial Relations and Technical Change in the British, Italian and German Automobile Industries: Three Case Studies'. Ed. W. Streeck. Wissenschaftszentrum Berlin: Discussion Paper IIM/LMP 85–5.

Streeck, W. 1985. 'Industrial Relations and Industrial Change in the Motor Industry: An International View'.

Streeck, W., P. Seglow, and P. Wallace. 1981. 'Competition and Monopoly in Interest Representation: A Comparative Analysis of Trade Union Structure in the Railway Industries of Great Britain and West Germany'. *Organization Studies*. Vol. 2, 307–29.

Thompson, J.D. 1967. *Organizations in Action*. New York: McGraw-Hill.

Trist, E. 1981. *The Evolution of Socio-Technical Systems. A Conceptual Framework and an Action Research Program*. Toronto: Ontario Ministry of Labour.

Warner, M. 1985. 'Microelectronics, Technical Change and Industrialised Economies'. *Industrial Relations Journal*. Vol. 16, 9–11.

Weick, K. 1979. *The Social Psychology of Organizing*. Reading, Mass.: Addison-Wesley.

Willman, P. 1985. 'The Implication of Process and Product Innovations for Labour Relations: A Review and some Research Technology', 28–9 May, 1985.

Woodward, J. 1965. *Industrial Organization: Theory and Practice.* Oxford: Oxford University Press.
Woodward, J. 1970. *Industrial Organization: Behaviour and Control.* Oxford: Oxford University Press.

2

Flexible Specialization: Miracle or Myth?

Richard Hyman

A new orthodoxy seems set to dominate the sociology of industry and organizations. A decade ago, against the background of the collapse of post-war capitalism's long expansionary phase, technological innovation and the reorganization of production were widely identified as a mechanism for the mass destruction of jobs and the degradation of many of those that remained. Yet today, despite the persistence of large-scale unemployment in the majority of Western economies, the prevailing mood of sociologists appears increasingly optimistic: microelectronic technology, it is commonly argued, provides the means not to reinforce but to reverse the methods established by Taylor and Ford, beneficiently transforming social relations within production.

Central to this analysis is the argument that the principles of 'scientific management' established in the early twentieth century were a contingent response to historically specific circumstances, not a universal and immutable expression of the logic of capitalist production. Particularly in the USA, commercial success could be achieved by capturing a mass market on the basis of a standardized product and competitive pricing. Product market strategy meshed neatly with a system of work organization based on machine pacing, fragmentation of tasks and tight labour discipline. On the basis of dramatic individual achievements, the 'mass production paradigm' (Piore and Sabel, 1984) became sanctified as a universal recipe, dominating economic institutions, investment policies, technological innovation, and patterns of industrial relations.

Changes in both product markets and labour markets, the argument continues, have made such methods a recipe for failure rather than success. In major part the economic crisis of the 1960s and 1970s was a crisis of mass production. The diversification of product ranges to meet more discriminating consumer demand, together with design alterations and product innovation, can be achieved without significant increase in unit costs – so it is claimed – by exploiting the adaptability inherent in computer-controlled production systems. The implications for work organization are considerable. 'Henry Ford's soul-destroying, wealth-creating assembly lines are out of date,' recently declared *The Economist* (5 April, 1986). 'Most of the things factories make now – be they cars, cameras or candlesticks – come in small batches designed to gratify fleeting market whims. The successful manufacturing countries in the twenty-first century will be those whose factories change their products fastest.'

The car industry, appropriately, provides the central point of reference for much of the recent literature on the obsolescence of Fordism. 'There are signs of a slow movement towards a model of production which turns the principles of mass production upside down. Instead of producing a standard car by means of highly specialized resources – workers with narrowly defined jobs and dedicated machines – the tendency is to produce specialized goods by general-purpose resources' (Katz and Sabel, 1985: 298). Traditionally it was regarded as axiomatic that no middle ground existed between the giant producer with a full product range, and the small producer catering for protected sub-markets; but today 'it is becoming easier for a medium-sized producer to use computer-aided design and flexible manufacturing to bring out variants of a few basic lines at relatively low cost' (Altshuler *et al*. 1984: 139–40). Modern technology permits innumerable product variations through different combinations of the same basic repertoire of components; the economies of standardization can be sustained even though 'on any day no two absolutely identical cars may roll off the line in final assembly' (Kern and Schumann, 1984: 43, my translation).

The claim that Fordist methods of labour management are now obsolete rests on the premise that 'flexible specialization' in product strategy both permits and requires analogous flexibility within the workforce. 'Mass markets are the precondition for the Fordist organization of production; when they begin to disintegrate, Fordism begins to lose its appeal within the factory. Where Fordism calls for the separation of conception from execution, the substitution of unskilled for skilled labour and special-purpose for universal machines . . . specialization often demands the reverse: collaboration between designers and skilled producers to make a variety of goods with general-purpose machines' (Sabel, 1982: 194). Employees must possess the readiness and ability to perform a variety of functions, and to acquire new competences as techniques and products evolve. Hence the thesis that new technology serves to increase rather than reduce skills; and hence also the concern to manage workers in ways which will stimulate their active commitment to efficient and high-quality production. Moreover advances in recording, processing and communicating data permit decentralization both within and between enterprises, again facilitating delegatory forms of management. This scenario thus presents as the model of future production, not River Rouge but the 'high-technology cottage industry' (Sabel, 1982: 220).

It cannot be denied that there are features in contemporary industrial developments which support this analysis. Among writers who embrace the notion of flexible specialization there are many who are cautious in positing a universal and irresistible trend. Nevertheless, in popular expositions of this thesis the vision of the factory of the future as a utopia where managements co-operate harmoniously with a polyvalent workforce is commonly presented with minimal qualification. A critical appraisal is therefore in order. In this chapter, four main issues are raised: the extent to which flexible specialization is, and can be, a coherent management strategy; the degree to which employer and worker interests can be harmonized within such a strategy; the 'other face' of flexibility as a source of insecurity and adversity for those outside the core labour force; and the broader social implications of new production strategies.

A New Managerial Solution?

The analysis of management strategy provides fertile ground for a rich blend of interpretation, prescription and *post hoc* rationalization. Among both radical and managerialist writers there is a tendency to view corporate decision-making as a planned and integrated process, with a clear and coherent articulation between staff and line, junior managers and top executives, administrative detail and general policies, short-term expedients and long-term objectives. This conception, it could be argued, entails an exaggeration of the unity and rationality of management; and a dangerous oversimplification of the strategic options available to employers.

The internal coherence of the management process cannnot be assumed *a priori*. The centrifugal tendencies of functional specialisms must be contained; the 'recalcitrance' of lower-level functionaries must be overcome. A substantial literature demonstrates that company decisions are commonly based on inadequate information (a problem which computerization will not automatically resolve); that even initiatives with substantial long-term consequences may be opportunistic and reactive rather than strategically directed; that careerism, empire-building and departmental politics are major elements in organizational dynamics (cf. Dalton, 1959; Burns, 1961). Even Chandler, the most notable popularizer of notions of employer strategy, has conceded

(1962: 11–12) that 'entrepreneurs . . . are not all necessarily imbued with a long-term strategic outlook. In many enterprises the executives responsible for resource allocation may very well concentrate on day-to-day operational affairs, giving little or no attention to changing markets, technology, sources of supply, and other factors affecting the long-term health of their company.' In Britain, certainly, it is a familiar comment that managements traditionally 'muddle through' in routine circumstances and respond to problems and crises with *ad hoc* reactions. In the specific case of the motor industry, a recent study comments that 'even on a modest definition of labour relations strategy, such as a set of related policies directed at problems which managements recognize as interrelated, the companies' industrial relations and manpower policies did not always constitute a strategy' (Marsden *et al.*, 1985: 35).

More fundamentally, strategic coherence may be obstructed not merely by the inadequacies and idiosyncrasies of managers as human actors, and by the recalcitrance of their informal organizational relations, but also by structural contradictions within the managerial process itself. Conflict and division within capitalist management reflect not merely the diverse ideologies and sectional interests of, say, marketing, production and personnel staff; different elements in the production and realization of surplus value may be *in principle* incompatible. 'For individual capitals – as for capital in general – there is no "one best way" of managing these contradictions, only different routes to partial failure. It is on this basis that managerial strategy can best be conceptualized: as the programmatic choice among alternatives none of which can prove satisfactory' (Hyman, 1987: 30).

Within specialisms too, inconsistency and fluctuation of policy may derive from contradictory pressures. The function of labour control, for example, involves both the direction, surveillance and discipline of subordinates whose enthusiastic commitment to corporate objectives cannot be taken for granted; and the mobilization of the discretion, initiative and diligence which coercive supervision, far from guaranteeing, is likely to destroy. As Burawoy puts it (1979: 30), the capitalist labour process 'must be understood in terms of the specific combinations of force and consent that elicit co-operation in the pursuit of profit'. It is of course a familiar argument that discretion is most likely to be permitted and encouraged among certain types of employee and in certain types of work context (Fox, 1974); in this sense, theorists of flexible specialization follow an established tradition in linking favoured employment status to the possession of polyvalent competences particularly valuable to management. Nevertheless there are few (if any) workers whose 'willing' co-operation is not advantageous to the employer; as Manwaring and Wood (1985) suggest in their discussion of 'tacit skills', the discretionary exercise of some form of special expertise can be found in virtually every labour process. But conversely, there are few (if any) workers whose voluntary commitment requires no external reinforcement. Within management itself, discretion can often be utilized to advance private rather than corporate interests (whether through 'white-collar criminality' or legitimate career advancement); conceding greater autonomy to workers (whether individually or collectively), particularly against a background of oppositional relationships, inevitably carries high risks.

Shifting fashions in labour management stem from this inherent contradiction: solutions to the problem of discipline aggravate the problem of consent, and *vice versa*. Accordingly, pragmatism may well be the most rational management principle. And it may indeed be questioned how far the elaborate and internally consistent programmes of 'management science' have ever guided actual management practice. There is evidence (e.g. Stark, 1980) that Taylorism, for example, was regarded with scepticism and suspicion by most practising managers; serving at most as a set of discrete recommendations from which to pick and choose, and thus negating Taylor's own prescriptions. Fordism itself, as Piore and Sabel argue (1984:69) 'was the end point of *ad hoc* innovations in production, not a deduction from a master idea'; nor did it subsequently serve as a universal model. In much of British manufacturing, certainly, what is notable is the long persistence of the mutually sustaining combination of 'bespoke' products, a craft or quasi-craft workforce, and *laissez-faire* management (Hyman and Elger, 1981; Littler, 1982; Samuel, 1977). What was widely denounced as the tyranny of 'restrictive practices' (or celebrated, on the left, as evidence of workers' control) may best be regarded as the historical product of a tradition of delegated, 'unscientific' management. Ironically, the shift towards a more Fordist model of product planning and labour control occurred in Britain most decisively as the new model of flexible specialization was allegedly taking shape elsewhere, in response to the crisis of the 1970s.

If, however, the coherence and universality of Taylorism and Fordism as managerial strategies are commonly exaggerated, the notion of a radically distinct alternative needs to be regarded sceptically. Again, actual management practice may involve the piecemeal application or rejection of elements of flexible specialization, rather than the adoption of an integrated package. How extensively, indeed, is such an integrated model applicable in principle? Most of the current literature focuses on manufacturing industry – which in Britain today employs only a quarter of the labour force. How far can the principles of post-Fordist product design and work organization be applied in the rest of the economy?

More fundamentally still: to pose flexible specialization as a potential recipe for *universal* corporate success is to envisage an economy in which there are no losers, only winners. Altshuler *et al.*, for example, propose the possibility of a 'new equilibrium' within which the less successful established car manufacturers recover their position, the new producers expand their markets, while the current world leaders hold their advantage (1984: 197). A delightful vision; but in a competitive capitalist economy not all contestants can win prizes. If some can gain a competitive edge through flexible specialization – and if self-contained market niches are not universally available – then some will be driven into the abyss.

A Harmony of Flexible Specialists?

For optimistic analysts, flexible specialization provides the route to a new interest accommodation between labour and capital; indeed, 'flexible specialization is predicated on collaboration' (Piore and Sabel, 1984: 278). Employers obtain a system of production best suited to commercial success; workers, through their willingness to move between jobs and train for new ones, enjoy more secure employment prospects within their company, as well as more challenging and responsible work. One question which at once arises is: *which* workers? The essence of this thesis is the consolidation of internal labour markets. Yet these are simultaneously mechanisms of inclusion and of exclusion; for those outside the havens of security, the obstacles to favourable employment status are made even greater.

This issue is considered further in the following section; meanwhile it is necessary to probe more critically the nature of developments within the internal labour market. It may first be noted that for many types of worker, strengthening the bonds of the internal labour market is an ambiguous gain: the fortress may also be a prison. The comparative advantage of the traditional craft worker was rooted in the *transferability* of scarce skills across the external labour market. Employer-specific training or job classification was resisted as a threat to mobility within a market in which the balance of supply and demand favoured the worker. In Britain at least, such resistance has proved effective until remarkably recently. It is clear, however, that increased flexibility in the internal division of labour reduces the general marketability of the worker's skills: external flexibility is undermined. In the short term, certainly, the consequence may be a mutual dependence of employer and worker; replacement costs of experienced workers may be high, precisely because substitutes cannot be obtained from the external labour market without further training. In the longer term, however, the relationship is asymmetrical. New products are developed and old lines are deleted; new factories are opened and old ones closed; new production systems are devised and obsolete installations scrapped.

Some firms simply go under, and with them their employees. What of the survivors? Whether or not ageing workers are willing to be retrained and relocated, there comes a point at which younger, cheaper, more flexible recruits represent a more advantageous investment. At this point an overt antagonism of interests emerges.

A second problem is to construct an accurate overall map of the impact of flexible specialization on job structures and job content. Company negotiators are keen to stress the transformation promised by 'flexibility agreements', less anxious perhaps to scrutinize and disseminate the actual results. Where product quality and reliability are important selling points, the quality and reliability of the workforce acquire marketing salience; it may be hard to separate advertising hype from shopfloor reality. Part of the difficulty is that detailed research has as yet covered a very limited range of contexts. Even accepting the authors' conclusions, questions of typicality inevitably arise. At best, the simplistic proposition that microelectronics lead universally to high-technology drudgery may be refuted; a more nuanced calculus of workers' gains and losses requires far broader evidence. The work of Coombs (1985), who seeks to differentiate three distinct 'regimes' of mechanization, each involving characteristic patterns of technical and organizational change, offers one suggestive though exploratory framework for more systematic analysis. Certainly the argument that microprocessor technology gives managements far greater potential for oppressive surveillance and control is in no way negated. 'Who writes the programs?' remains the crucial question. So long as the software is designed 'to control labour rather than emancipate it' (Murray 1985: 31), it would be naive to assume that conflict over this issue can be permanently avoided.

Thirdly, there are uncomfortable parallels between some of the current rhetoric on 'flexible specialization' and the traditional literature on 'participation' and 'job enrichment'. Discussion is often pitched at a level of generality which inhibits analytical and conceptual precision. If – as some argue – all forms of work involve 'tacit skills', what are the terms on which the deskilling/reskilling debate can meaningfully be conducted? The delegation to the immediate work group of detailed decisions on organization and scheduling of production, and responsibility for initial monitoring of product quality, does not negate the overall directive control of management: a control which can be applied coercively if self-disciplining fails to yield the required outcomes. 'Part of the attraction of job reform, as distinct from some other forms of participation, is that it is "soft on power", affording little or nothing in the way of concessions on business decision-making and carrying little danger of getting out of hand' (Ramsay, 1985: 74–5). Delegated management does not equal self-management; nor does an expanded portfolio of competences neccessarily equal enhanced skill. Employees who can perform a variety of routine functions will doubtless be of greater value to the employer; but their own sense of meaning, dignity and control in work may reveal no significant improvement. In the much-quoted words of a chemical worker (Nichols and Beynon, 1977: 16): 'I never feel "enriched" – I just feel knackered'.

Finally it is necessary to consider the institutional framework of collaboration. Kelly (1985: 42) argues that 'a "successful" job redesign scheme is one which has been modified through negotiation so that both labour and capital derive benefits as well as sharing costs'. Managements seek improved quality, rapid adaptability to changing production requirements, and also reduced labour costs. The objective compatibility of these goals with workers, expectations and aspirations will vary, inevitably, according to context. So will *perceptions* of the balance of common and conflicting interests. Altshuler *et al.* (1984: 217–18) note that a collaborative restructuring strategy is most likely to be pursued by managements 'where there is a tradition of institutionalized co-operation'. Elsewhere, 'managers may feel themselves caught in the vicious spiral . . . of declining legitimacy, tightening discipline and collapsing authority. But this does not mean they may simply reverse the process and start a virtuous circle of regaining control by sharing it.. . . The productivity released by encouraging the creativity, initiative and responsibility of the workforce may be great, but there is no reason why it should serve the interests or purposes of management' (Cressey *et al.*, 1985: 172). An adversarial tradition of industrial relations inevitably increases the salience of opposing interests in the distribution of costs and benefits. And objectively, the relative weight of costs necessarily increases with the seriousness of management's competitive predicament. Recent British experience certainly suggests a sharp trend *away* from compromise and accommodation. And if the level of overt industrial conflict has actually declined, this might better be taken as evidence of labour's weakness and demoralization than of a new co-operative spirit.

Flexibility as Insecurity: the Costs of Exclusion

As indicated above, flexible specialization has important implications for the structure of labour markets. The aim is to consolidate a stable core of employees able to adapt to both cyclical and secular changes in the level and composition of production. In Atkinson's phrase (1985: 11), such a reorganization of the internal division of labour is a means to *functional flexibility*; and those who are able to guarantee management the desired adaptability may possibly benefit in terms of job security and conditions of employment.

But firms in unstable product markets are also concerned to obtain *numerical flexibility*, and here the implications are altogether different. The central objective is to render workers disposable rather than adaptable. Among the means adopted are the recruitment of workers on short-term contracts; the use of subcontractors rather than direct employees; resort to hire-and-fire policies; employment of part-time workers who fall outside job protection legislation. Within many companies there is evidence of a growing dichotomy between core and peripheral workers. The same trend is also apparent between companies: those firms acting as contractors to dominant firms providing particularly insecure employment (as in the Japanese model?). In Britain, the 'privatization' of many processes in state institutions (e.g. cleaning, catering and laundry work in hospitals) is a parallel instance (Hastings and Levie, 1983). The latter case also indicates a more general point: that the emphasis on manufacturing in the literature on flexible specialization may in this respect be particularly misleading. In 'service' industries, arguably, peripheral status is increasingly the norm; firms compete primarily by economizing, largely at the expense of the workforce (Atkinson and Gregory, 1986).

One variant of this trend has been analysed by Mitter (1986) in a recent study of the UK clothing industry. Just as the international rise of Benetton as a fashion giant rests upon low-paid female domestic toil, so oligopolistic rivalry among the household names of Britain's high streets is sustained by a network of competing small suppliers whose own viability depends on the sweated labour of Asian and Cypriot women immigrants. Successful adaptability to fashion trends is thus rooted in the vulnerability of the most insecure and disadvantaged sections of the workforce.

Flexibility thus entails intensified segmentation within the workforce, between the relatively sheltered and advantaged, and the vulnerable and oppressed. For the latter, as the European Trade Union Institute has argued (1985:2–3), flexibility normally means 'cutting real wages, cutting lowest wages most; breaking up national negotiating procedures; abolishing employment legislation; making it easier to sack people; increasing job insecurity; attacking social security systems; and dismantling health and safety and environmental protection'. And, one should add, obstructing union organization and collective action. Peripheral status is of course a reflection and reinforcement of other social divisions: by gender, ethnic origin, age and education. Geographical inequalities are also reinforced. In Britain, for example, it has been argued (Massey 1983) that there is a growing dichotomy between the 'sunbelt' (primarily south-east England) with highly paid and high-status work in senior corporate management, research and development and advanced technology production; and the rest of the country marked either by industrial decline or by the development of low-status, disposable production and clerical operations.

Cross-nationally an analogous differentiation is apparent. While some formulations of a 'new international division of labour' are oversimplified, there is considerable evidence of a restructuring of transnational capital which reinforces core–periphery divisions between countries. Advances in computerization and telecommunications facilitate the concentration of 'conception' (research, planning, directive and strategic management) at corporate headquarters, while 'execution' is dispersed around the globe in often low-cost installations which may be abandoned with market alterations. Microprocessor technology is itself a prime example: the production of basic microelectronic components depends considerably on the labour of young women in Third World 'free trade zones'. *The Second Industrial Divide* is a title rich with irony.

Flexibility and Fetishism

Rationalization within individual productive units, anarchy within the national and global economy: current trends reconstitute the familiar contradiction between micro- and macro-efficiency within capitalist production. Any assessment of the social implications of flexible specialization must, however, also take account of the relationship between production and consumption: an issue which transcends most conventional conceptions of industrial democracy, but which trade unionists have not altogether neglected (see, for example, Collective Design, 1985).

However partially and one-sidedly, the mass-production/mass-consumption model expressed an obvious 'democratic' principle. In standardized form, items which were already established as desirable but attainable only by the rich became generally available. Fordism's populist aspect was no doubt one reason why many socialists acclaimed its progressive character even while denouncing Ford's specific employment practices. While the accompanying egalitarian rhetoric must be regarded sceptically, the orientation of production to mass markets clearly did entail levelling tendencies within the sphere of commodity consumption.

Against this background, the recomposition of commodities which forms the product-market background to flexible specialization should be viewed as a regressive trend. Three separate features may be noted. Firstly, an often spurious attempt at product differentiation is designed to symbolize consumer *in*equality: 'a relegitimation of luxury consumption by the ruling class' (Gough, 1986:63). The sales strategy is to emphasize that the product is not merely different but superior, and that this superiority is transferred to the purchaser: the commodity conditions the social existence of the consumer. 'The reproduction of inequality – poverty and privilege – at ever-higher levels is a necessary condition for the indefinite growth of demand,' argues Gorz (1985:22). 'It feeds on – and is fed by – an ideology of privilege in which what is good enough for everyone is not good enough for anyone.' Such a competitive pursuit for superiority, as Hirsch (1977) has elegantly demonstrated, is both socially and economically destabilizing.

Secondly, demand is manipulated by the propagation of new 'needs': generating the novelty–obsolescence nexus, and hence ensuring a market for new or altered products regardless of value in use. While the dominance of marketing over product design and production organization (a hegemony cloaked in the slogans of consumer sovereignty) is no new feature in modern capitalism, its salience would seem to be significantly increased. Optimistic analysts of 'flexible specialization', by contrast, typically if naively assume the primacy of production within the circuit of capital. Yet the prospect of an economy beneficently inspired by progressive engineers – the vision which exerted widespread influence in both America and Europe around 1920 – should be received with greater scepticism today.

Thirdly, an increasing proportion of social production and social consumption assumes the form of marketed commodities; both as producers and consumers, in an expanding area of our lives we are engaged in a relationship with capital. Commodity fetishism extends its grip. As Gorz has argued (1985:49)

> no longer is work a way of living and acting together, no longer is the workplace a place of life, work time a reflection of seasonal and biological rhythms. Money, in the form of profits or wages, is the overriding goal of every activity, rather than pleasure or satisfaction. The triumph of market relations over relations of reciprocity, of exchange value over use value, has impoverished our lives and abilities.

Again, any analysis which concentrates on relations in production to the neglect of relations of exchange fails to confront such tendencies.

Conclusion

The intensification of commodity relations, and of wage labour as an integral component of these relations, proceeds at the very moment when technological innovation could enable a relaxation of the grip of traditional employment structures; and when growing numbers are in any event being expelled from stable, regular, full-time waged work. An effective trade union response to flexible specialization must thus transcend the immediate implications for work relations within the enterprise. Macroeconomic and macrosocial considerations are essential items on the agenda. Among these considerations, the *what* and the *why* as well as the *how* of production decisions become of key relevance. The labour movement's old demand that production should be determined by social need rather than profit assumes new meaning and new urgency with the advance of microelectronic technology. Flexible specialization can constitute a humane development only if liberated from the dominance of capital; only if it serves to loosen, not reinforce, the hold of commodity production and the wage–labour relationship (see, for example, Gershuny, 1983; Gorz, 1982). To argue thus is to insist on the contested status of flexibility as a concept. To redefine flexible specialization – to distinguish essence from rhetoric – is at the same time to give richer meaning to that other contested concept, industrial democracy.

References

Atkinson, J. 1985. *Flexibility, Uncertainty and Manpower Management.* Brighton: IMS Report 89.

Atkinson, J., and D. Gregory. 1986. 'A Flexible Future'. *Marxism Today.* April.

Altshuler, A., M. Anderson, D. Jones, D. Roos, and J. Womack. 1984. *The Future of the Automobile.* London: Allen & Unwin.

Burawoy, M. 1979. *Manufacturing Consent.* Chicago: University of Chicago Press.

Burns, T. 1961. 'Micropolitics'. *Administrative Science Quarterly.* Vol. 6.

Chandler, A.D. 1962. *Strategy and Structure.* Cambridge, Mass.: MIT Press.

Collective Design. 1985. *Very Nice Work If You Can Get It.* Nottingham: Spokesman.

Coombs, R. 1985. 'Automation, Management Strategies and Labour-Process Change'. *Job Redesign.* Ed. D. Knights, H. Willmott, and D. Collinson. Aldershot: Gower.

Cressey, P., J. Eldridge, and J. MacInnes. 1985. *Just Managing.* Milton Keynes: Open University Press.

Dalton, M. 1959. *Men Who Manage.* New York: Wiley.

European Trade Union Institute. 1985. *Flexibility and Jobs.* Brussels: ETUI.

Fox, A. 1974. *Beyond Contract.* London: Faber.

Gershuny, J. 1983. *Social Innovation and the Division of Labour.* Oxford: Oxford University Press.

Gorz, A. 1982. *Farewell to the Working Class.* London: Pluto.

Gorz, A. 1985. *Paths to Paradise.* London: Pluto.

Gough, J. 1986. 'Industrial Policy and Socialist Strategy'. *Capital and Class.* Vol. 29.

Hastings, S., and H. Levie. 1983. *Privatisation?* Nottingham: Spokesman.

Hirsch, F. 1977. *Social Limits to Growth.* London: Routledge & Kegan Paul.

Hyman, R. 1987. 'Strategy or Structure? Capital, Labour and Control'. *Work, Employment and Society.* Vol. 1.

Hyman, R., and T. Elger. 1981. 'Job Controls, The Employers' Offensive and Alternative Strategies'. *Capital and Class'*. Vol. 15.

Katz, H., and C. Sabel. 1985. 'Industrial Relations and Industrial Adjustment in the Car Industry'. *Industrial Relations*. Vol. 24.

Kelly, J. 1985. 'Management's Redesign of Work'. *Job Redesign*. Ed. D. Knights, H. Willmott, and D. Collinson. Aldershot: Gower.

Kern, H., and M. Schumann. 1984. *Das Ende der Arbeitsteilung?* Munich: Beck.

Littler, C. 1982. *The Development of the Labour Process in Capitalist Societies*. London: Heinemann.

Manwaring, T., and S. Wood. 1985. 'The Ghost in the Labour Process'. *Job Redesign*. Ed. D. Knights, H. Willmott, and D. Collinson. Aldershot: Gower.

Marsden, D., T. Morris, P. Willman and S. Wood. 1985. *The Car Industry: Labour Relations and Industrial Adjustment*. London: Tavistock.

Massey, D. 1983. 'The Shape of Things to Come'. *Marxism Today*. April.

Mitter, S. 1986. 'Industrial Restructuring and Manufacturing Homework'. *Capital and Class*. Vol. 27.

Murray, R. 1985. 'Benetton Britain'. *Marxism Today*. November.

Nichols, T., and H. Beynon. 1977. *Living With Capitalism*. London: Routledge & Kegan Paul.

Piore, M., and C. Sabel. 1984. *The Second Industrial Divide*. New York: Basic Books.

Ramsay, H. 1985. 'What is Participation For?' *Job Redesign*. Ed. D. Knights, H. Willmott and D. Collinson. Aldershot: Gower.

Sabel. C. 1982. *Work and Politics*. Cambridge: Cambridge University Press.

Samuel, R. 1977. 'The Workshop of the World'. *History Workshop*. Vol. 3.

Stark, D. 1980. 'Class Struggle and the Transformation of the Labor Process'. *Theory and Society*. Vol. 9.

3

Comparative Research and New Technology: Modernization in Three Industrial Relations Systems

Beat Hotz-Hart

Substantial changes are now under way in the structure of industrial society. Companies are under extreme challenge. They are forced to react to major changes such as fast moving developments in technology (particularly the introduction of microelectronics), the increasing internationalization of capital and the international division of labour. Piore and Sabel (1984) highlight the importance of these changes in speaking of a 'second industrial divide'. Their central hypothesis is that industrial economies are in the midst of a shift between two modes of technical development: mass production and flexible specialization. They postulate that for most companies in highly industrialized countries this means structural change either towards specialization, flexibility and permanent innovation with skilled workers for market niches, or towards a new Fordism: capital-intensive production with robots for large, stable and growing markets. The basis of an explanation of what happens in industry, however, does not concern solely the application of new technology but *how* decisions are taken in this respect, and *for what purpose*. Therefore, at this point in economic history it is the relation between economic performance and institutional mechanisms to which increasing attention has to be paid. This raises the question, what is the role of institutions of employment and labour in economic performance and social progress, and how do such institutions adapt to changes and encourage or restrict them?

Within this broad question, I have focused on the impact – if any – of certain characteristics of industrial relations prevalent in a country on the strategies and behaviour of companies in managing the restructuring process and, therefore, on long-term company development. This chapter summarizes the method of analysis and some of the results from an extensive study I have undertaken in this field (Hotz-Hart, 1987). My analysis focuses on some aspects of the complex configuration of constraints and opportunities created by industrial relations institutions for managerial strategic choice, particularly in connection with the challenge presented by new technology. How do industrial relations influence the strategies and capabilities of enterprises with regard to the introduction of new techniques, and how are they in turn changed by those new techniques?

Before outlining the main argument, some general questions of method should be addressed. Research involves the application of models, explicit or implicit, quantitative or qualitative. In cases where the impact of variables such as institutional aspects of industrial relations is under study, the variables dealt with are stable within one industry or country over a long time period: they will be seen as exogenous. If we wish to assess the impact of a variation of these exogenous variables we have to find ways to apply our model to cases where these variables are different, and as many other things as possible are equal.

It is possible to try to answer the questions raised by looking at different industries within one country which display different systems of industrial relations. But such an approach is limited because the variation of the institutions might not be wide enough, or the technological background of the different industries might differ too much. A more attractive alternative is, therefore, to look at the same industry, with similar products and production techniques, and similar challenges from technological developments, operating in different institutional settings across different countries. I have chosen to compare the engineering industries of West Germany, Britain and the USA. In doing so, several problems arise: in order to apply a model, adequate cases or sectors within different national economies have to be identified and separated. Therefore, a crucial and difficult step in the analysis is the identification of a set of variables according to the industry's problem of coping with new technology which form a functional entity and can reasonably be isolated and compared. This requires for each country an in-depth understanding of highly idiosyncratic institutions performing a complex variety of functions.

The set of variables chosen has to be functionally connected with the problem of company behaviour under the challenge of new technology and the labour–management relationship. This requires the identification of 'functionally equivalent mechanisms' between the countries which can be analysed and compared. These functions have to be seen in the overall context of the countries' industrial relations systems as a whole. Often, large-scale cross-section analysis of some few variables from national statistics misses important points. But a pure description of cases in similar areas is not sufficient either. There is 'the danger of lapsing into either vacuous description or superficial comparison' (Shalev, 1980: 40).

International comparative research needs a strong theoretical framework on an adequate level of complexity and abstraction and precisely here lies a crucial problem. The state of the art is strong in fact-finding but lacking in generally applicable theories. A good deal of the existing work within the comparative industrial relations field, although often rich and insightful, has not so far been explicitly theoretical in either its purpose or method, a view which is supported by Shalev (1980). Comparative studies can begin with an *a priori* hypothesis about the relationship between industrial relations arrangements and company strategies of coping with new technology, and proceed to test their validity via an examination of the experience of a number of different countries. It is useful to study the countries in terms of pre-established categories drawn from a preliminary analysis of discussions in this field. Empirical work will lead to a revision of the model. Finally, as Bain and Clegg (1974) stress, there has to be a reciprocal, feedback relationship between theorizing and observation.

Links between Industrial Relations and Company Strategies

Figure 3.1 summarizes the structure of the research. Each of the categories is outlined in more detail below. In comparing the engineering industry of West Germany, Britain and the United States five dimensions of industrial relations within a company are highlighted: (1) job control: who has it and at what level does it operate? (2) the structure of qualifications and jobs within the company; (3) organizational modes of labour; (4) the pattern of interest representation; and (5) conflict regulation and the degree of formal regulation within industrial relations. In reviewing the literature these characteristics were seen as the most important, and the most likely to have an impact on a company's ability to change.

At the time of the research Britain had a decentralized, relatively high degree of job control by workers, job territories and demarcations, shop stewards and multi-unionism, a multiplicity of labour organizations, direct interest representation through shop stewards, a fragmented collective bargaining system, a low level of centralization, a relatively under-institutionalized and only minimally legalized system of regulation.

Industrial relations in Germany are characterized by a high level of centralization, legal recognition and institutionalization, dual structure of collective representation with a division of labour between work council and unions, a limited and equally distributed means of interest

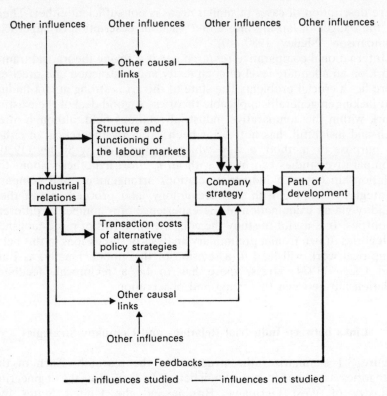

FIGURE 3.1
The Impact of Industrial Relations on the Development of Companies
and Industries

representation, the concentration of internal and external authority at the top of the unions.

In the United States industrial relations are characterized by decentralized job control by workers through job rules, scientific management, highly firm-oriented training, indirect interest representation, highly professionalized and bureaucratic, national and local union organization, a conflictual relationship and detailed contracts which are substantively free but procedurally legally regulated.

Two issues for developing a theoretical argument for explaining a possible and assumed causal link between industrial relations and structural change in companies are suggested. It is assumed that, firstly, there is a causal relationship between industrial relations and the structure and functioning of the labour market (particularly the internal labour market) which in turn has an impact on company strategies and development. Secondly, it is argued that there is also a link between industrial relations and the size and structure of transaction costs of companies in the modernizing process which in turn have an impact on company development. For analytical purposes these two aspects were studied separately, although they are not independent of each other.

The Structure and Functioning of the Labour Market
The activities of workers at the level of the enterprise or an industry are an active factor which has a formative and modifying effect on the structure and functioning of the labour market. Simultaneously, management pursues personnel and organizational strategies which also shape the labour market. These pressures and countervailing pressures, the negotiation processes involved, take place within a given industrial relations setting. Therefore it is assumed that characteristics of industrial relations as outlined shape the structure and functioning of the labour market. The analysis concentrated on the shaping of the following labour market aspects in order to compare the engineering industries of the three countries in that respect: (1) recruiting patterns; (2) the possibility of substitution or reallocation of labour; (3) the functioning of the internal labour market, i.e. flexibility of the shopfloor; (4) qualification process; (5) dismissals and labour turnover.

Empirical analysis along these dimensions showed that there are substantial differences in the functioning of the labour market between the countries particularly with respect to flexibility on the shopfloor caused amongst other things by institutional factors. Within US companies the internal labour market is highly developed but, through seniority rules, structured in detail. It is therefore rather rigid with little freedom for management to reallocate labour. As a consequence employment adjustments to economic change are realized through 'lay-offs' according to 'last in first out', and high turnover rates. There is little interest in training and retraining on both sides.

In German companies the internal labour market is highly developed too but less structured and therefore more flexible than in the US. Management has more right to reallocate labour. As an outcome of industrial relations there is a substantial amount of internal employment policy and stability and there are strong efforts made towards training and retraining on both sides. The formation of an internal labour market offers the members of the permanent staff relatively high job security but also ties them quite closely to the enterprise. Both aspects can be seen in close connection with German industrial relations.

British companies seem to have the strongest craft orientation amongst the three countries. Internal labour markets are highly fragmented and strictly divided into 'job territories': there is little internal flexibility. External employment adjustments take place only under strong pressure from outside. Training and retraining have been neglected for a long time.

It is argued later that these labour market characteristic become more and more important for company development if the company or the industry is confronted with structural change.

The Transaction Costs of Companies in the Modernizing Process

In a second argument, structural change is seen as a *transaction* problem in line with the argument of Williamson (1979) and others. To pursue any kind of strategy in order to cope with structural change, certain transaction costs have to be taken into account by management and labour. It is assumed that the size and structure of these transaction costs and their distribution amongst management and labour depend on the industrial relations systems within which the company operates. Similar strategies may involve substantial differences in the amount and distribution of transaction costs depending on the characteristics of industrial relations.

In order to discuss and assess the cost differences, indicators which can be operationalized and identified in the countries under review had to be found. Based on transaction cost theories the following ones were used: (1) existing obligations through labour law; (2) the content of collective agreements and the degree of contractual regulation; (3) procedural costs such as personnel involved in interest representation; (4) number of interest organizations involved (multi-unionism); (5) degree of legal regulation of the procedures; (6) conflict intensity (strikes); (7) the role of unions in implementing the agreements and new policies.

German companies have to cope with higher transaction costs stemming from the legal obligations associated with particular strategies and from procedural matters including high personnel involvement on both sides. On the other side, the content of collective agreements is less complicated and detailed than in the US, fewer organizations are involved, conflict intensity is lower (there is a flexible approach to problem-solving, an ability to absorb conflicts and a readiness to compromise) and unions participate in implementing agreements to a larger extent than they do in other countries.

The US engineering companies are characterized by quite contrary aspects: management enjoys an advantage through fewer legal constraints and smaller procedural costs. The disadvantages consist mainly in more intensive conflicts between management and labour, and the fact that the solutions are more detailed and collective agreements complex. Unions provide little support for policy implementation.

British companies seem to be in a disadvantageous position with respect to all cost indicators except one: the degree of legal regulation of procedures. Compared with Germany and the US there is no aspect

TABLE 3.1
Possible Company Strategies

Possible Company Strategies	The Resulting Path of Company Development
1. Innovative dynamic strategy Active and initiating renewal Imitation	Renewal, rejuvenation
2. Consolidation of well-established positions Rationalization and reduction or selective reduction of activities Intensification	Ageing

which is obviously less advantageous in the other countries. The level of transaction costs which goes with modernizing processes is the highest of the three.

Alternative Company Strategies

Institutionalized relations between workers and management affect the speed of technical change and its quality and direction. After comparing the functioning of the labour market and the transaction costs caused by industrial relations characteristics, it is possible to identify distinctive company strategies in dealing with challenges caused by new technology. The options shown in table 3.1 may be distinguished.

The research reveals more progressive strategies of innovation and restructuring and more conservative strategies of work intensification and cost reductions through rationalization. Four areas of these company strategies which could be observed in the countries deserve particular attention: personnel, organization development, competitive and investment behaviour.

Each strategic alternative mentioned in table 3.1 is characterized by a certain shape of these policy variables which together form a particular interdependent pattern (for two idealized types, see table 3.2). If the company intends to pursue a particular strategy these company variables as a whole have to be moulded or developed in an adequate way and implemented. Management and labour represent different interests in assessing the possible strategy alternatives and the development of the different policy variables connected with them. Conflicts arise, for example, over the distribution of costs of adjustments and over the earnings of a successfully implemented new strategy. According to the

TABLE 3.2

The Shape of Company Policy Variables with Respect to Overall Company Strategy (Ideal–Type Cases)

Company Strategies	(1) Innovative Dynamic	(2) Consolidation of Well Established Positions
Policy Variables		
Personnel	Activating and developing internal and external means of improving qualifications; increasing internal mobility; avoiding lay-offs if possible in a downturn	Improving efficiency through work intensification and reduction of labour costs; little training or retraining for new jobs; lay-offs are routine when business declines
Organization development	Increasing flexibility and adaptability; development towards co-operative and participative structure; teamwork; decentralization; suggestions from bottom up	Improving control and adjusting layout of work-places; authoritarian, patriarchal leadership; maintaining and reinforcing centralization; top-down command structure
Competitive behaviour	Aggressive; dynamic through product and process innovation; new ventures	Protective; market closure; reinforcing price competition; streamlining to the most profitable parts
Investment behaviour	In innovation, accepting higher risks, less capital intensive	Cost optimization; rationalization; capital intensification

theoretical framework applied here the strategy choice and its implementation through the variables distinguished has to be seen within the structure and functioning of the labour market and the level and distribution of transaction costs.

Company Development

Next the impact of industrial relations on company policy has to be assessed, which raises new methodological problems. A strong 'elective affinity' (Weber) is assumed between requirements of possible company strategies of industrial adjustment and a 'cluster' of industrial relations characteristics as outlined above. Where one specific characteristic of strategies and/or industrial relations is present, the others tend to be also present, and where one is absent, the others are likely to be absent as well (see Streeck, 1985; 15, 22). The variables characterizing alternative policy strategies were compared with the 'structure and functioning of the labour market' and with the 'level and distribution of transaction costs' both of which are connected with particularities of industrial relations. It was discussed whether the market and cost patterns of engineering in each country supported or inhibited a particular policy strategy. The consistency or inconsistency of the policy pattern and the labour market and cost pattern were checked. Conclusions were drawn with respect to the incentives and disincentives for the choice of a particular policy alternative. As a result conclusions could also be drawn with respect to the most likely (longer-term) company or industry development under these conditions. The nature of comparative research allows only statements about the *relative* position of the countries, not an 'absolute' assessment.

The results of the analysis indicate that industrial relations does affect the choice of company policy and therefore the pattern of company development. The same challenge stemming from the economic environment is met in substantially different ways by engineering companies in the different countries. Within a different industrial relations context the same technology is likely to be used in a different way.

As is outlined below, crucial characteristics of industrial relations in the US and UK engineering industries are not functional for an innovative strategy. They delay and hinder numerous changes and adjustments which go together with such a policy pattern. Differences between US and UK companies are not obvious. whereas German companies are in a comparatively advantageous position.

British industrial relations give incentives to and foster company strategies of rationalization and work intensification or of streamlining, shrinking, selective reduction of activities, or closures. In the long run British industrial relations give strong incentives to a development which is characterized by ageing and consolidation and not by innovation and renewal.

In the US engineering industry – as in Britain – rationalization and work intensification, a 'new Fordism', are stimulated by existing industrial relations. Incentives towards ageing as a long-term trend are prevalent too.

Industrial relations in the German engineering industry, with respect to the functioning of the labour market, provide far better conditions for a strategy of innovative dynamic and renewal. On the other side,

..re higher procedural costs and legal constraints which delay
.estructuring in the German industry. But all in all German industrial
relations do not constrain the pursuit of innovation as a company
strategy to the same extent as in the other two countries.

To sum up, British industrial relations present clearly the most
disadvantageous conditions for companies which attempt to cope with
structural change through renewal. German industrial relations cause
transaction costs above the international average. But these are no
serious disadvantage as long as the other industries are substantially
hindered by their own industrial relations. The unionized US engineering industry is somewhere in between, probably closer to Britain than
to Germany.

Feedback Processes

It is widely accepted that structural change, particularly flexibility and
innovation, has become necessary for the survival of most industries
in advanced countries. Long-term profitability can be pursued only to
a limited extent through cost minimization, rationalization and work
intensification. But flexibility and innovation are precisely the strategies
which are highly constrained by many industrial relations characteristics
of the (unionized) engineering industry, particularly in Britain and the
US. Therefore these industrial relations characteristics are substantially
challenged by employers. Adjusting industrial relations and labour
market structures becomes itself part of company strategy, although
implementing change might need quite some time. In the long run, a
feedback process takes place (see figure 3.1). The evolution of an
industry has to be seen as a dialectical relationship between (industrial
relations) *structure* and (decision-making) *processes* by various actors.

Therefore the questions at the beginning of this chapter may be
turned around: what consequences do technological change and its
repercussions on labour market structure through company strategies
have for industrial relations and trade union policy? These consequences
become very serious with change in institutional characteristics which
form part of the basis of the existence of the unions. A company's use
of new technology may undermine existing institutions and practices
of industrial relations and lead to the emergence of new ones. How
can unions meet these changes? What are the methods and ways of
coping with institutional change?

This brings us back to a methodological point: Comparative research
has to deal explicitly with the *historical* dimension, or as Hyman (1982:
108) puts it: in industrial relations 'the present must be viewed as
historically conditioned and historically contingent, and as incomprehen-
sible except by means of historical understanding'. Therefore 'inter-

national differences cannot be understood solely in ter/
sectional analysis at any one point in time. Instead, longituαιι.
incorporating a time dimension are also required for supplying histo..
perspective, together with a sharper appreciation of change through
time and the conditions which generate it' (Bean, 1985: 11).

But including the time dimension in comparative analysis is not
without problems. The example of Kerr *et al.* (1971) shows that in the
literature there is a tendency towards a 'convergence or evolutionist
theorizing' either to a particular point or more advanced to a range of
alternatives. Different stages of development that societies have reached
along some continuum are identified. It has to be shown where a
particular industry is located and where it is likely to go.

A prominent example is an argument of Olson's seminal work *The
Rise and Decline of Nations* (1982). He assumes a more or less
deterministic evolution of societies towards the development of a
collusion amongst special interest organizations, a 'gradual accumulation
of distributional coalitions' (79). Distributional coalitions interfere with
an economy's capacity to adapt to change and to generate innovations,
and therefore reduce the rate of growth. Societies are seen as being
in different phases of such an evolution.

The economic growth of a nation depends on its place in this process.
'The logic of the argument implies that countries that have had
democratic freedom of organization without upheaval or invasion the
longest will suffer the most from growth-repressing organizations and
combinations' (77). Olson's theory 'predicts that with continued stability
the Germans and Japanese will accumulate more distributional
coalitions, which will have an adverse influence on their growth rates'
(76).

In his argument Olson interprets institutions as constraints on market
competition, but presents no positive argument about how industrial
change and renewal take place, and what impact differences in
institutional settings (such as industrial relations) cause except in
assessing these in terms of their influence on market structure. In this
respect my analysis shows that it is not only the age of the institutions
or their existence or absence which explains the character of industrial
development. The argument has also to take into account their
characteristics and functions. The impact of institutional settings can
then be understood as dependent on the historical phase and the larger
context. Some settings which are viewed today as having a retarding
impact, were considered earlier as positive for industrial development,
and institutions which are today accused of detrimental effects may in
the future be regarded as beneficial.

Within certain limits industrial development is contingent and cannot
be understood as a deterministic or mechanistic process – which is not
to say that structurally similar responses cannot arise from widely

different cultural and historical contexts. A realistic discussion might settle somewhere between the rather deterministic view of Olson and the rather idealistic view of rational choice by Piore and Sabel.

The study presented here analyses the unionized engineering industry in different countries and argues that industrial relations aspects make the adjustment strategy in these industries in the US and UK much more likely to be 'new Fordism' than flexible specialization and new forms of production such as autonomous groups. German industrial relations seem to offer better chances for such a development.

A recent management strategy is to bypass the existing constraints of industrial relations. Particular strategies within individual companies are pursued, with or without unions, or by building up new types of worker representation, which pay little heed to, or substantially amend, the established order of industrial relations. In German industry the contractual and legal system allows little bypassing, while such a strategy is very realistic for US companies or plants and is obviously pursued in many cases. To this extent US companies are in a favourable position. 'Bypassing strategies' are possible in the UK too, but more difficult than in the US. Changes in this respect will increase the pressure on the Germans to amend their industrial relations.

Together with the changes in the international division of labour goes an increasing international competition between the existing institutional arrangements within each industry. The challenge for the development of the industries is transferred to a growing pressure to develop new industrial relations structures, and that depends on the social capacity for institutional change. There is a growing challenge for the unions to develop adequately their own institutions and their strategies of interest representation if they want to survive. In the long run unions have to realize substantial social innovations or they will be reduced drastically in their influence or even replaced. The traditional form of labour movement will not be able to survive for long. Clearly the employment relationship will be reshaped substantially with or without unions.

But in real life, there is a considerable degree of inertia in industrial relations systems although it differs widely across countries with similar technological bases. In Britain the general inability to innovate is also evident in industrial relations. There is extraordinary resistance to attempts at change. Although there has been a 'dissolution of outdated forms of organisation and representation', 'no politically, economically and socially viable transformation model for a new stage in social development has yet emerged' (Jacobi *et al.*, 1986: 6).

More recent developments seem to show at least some convergence among industrial relations structures towards company or plant unions. This is likely to include forms of co-operation between employees and employers at the plant and company level which are functional even without the formal organization and existence of unions and that would again be 'bypassing' the established industrial relations. There is a tendency that the employment relationship is reshaped without unions being involved. More recent US developments give evidence of this. The German line of discussion concerning institutional change is to develop models of co-determination focusing on issues of technological change.

Unions have to become involved in designing production strategies at company level and allow at the same time flexibility on the shopfloor. They have to do significantly more than negotiating on distributional matters. But what does it really mean when labour takes over 'some' control? What does industrial democracy really change compared with the influence of the 'laws' of capital accumulation and competition? Here, a large field for research and policy suggestions is open.

Note

I would like to thank Moira Hotz-Hart for helpful comments and support.

References

Bain, G.S., and H.A. Clegg. 1974. 'A Strategy for Industrial Relations Research in Great Britain'. *British Journal of Industrial Relations*. Vol. XII.

Bean, R. 1985. *Comparative Industrial Relations*. London: Croom Helm.

Hotz-Hart, B. 1987. *Modernisierung von Unternehmen und Industrien bei unterschiedlichen industriellen Beziehungen*. Bern, Stuttgart: Haupt.

Hyman, R. 1982. 'Review Symposium'. *Industrial Relations*. Vol. 21.

Jacobi, I.O., H. Kastendiek, and B. Jessop. 1986. 'Between Erosion and Transformation: Industrial Relations Systems Under the Impact of Technological Change'.*Technological Change, Rationalisation and Industrial Relations*. Ed. O. Jacobi., H. Kastendiek, and B. Jessop. London: Croom Helm.

Kerr, C., J.T. Dunlop, F.H. Harbison, and C.A. Myers. 1971. 'Postscript to Industrialism and Industrial Man'. *International Labour Review*. Vol. 103.

Olson, M. 1982. *The Rise and Decline of Nations*. New Haven, London: Yale University Press.

Piore, M.J., and C. Sabel. 1984. *The Second Industrial Divide*. New York: Basic Books.

Shalev, M. 1980. 'Industrial Relations Theory and the Comparative Study of Industrial Relations and Industrial Conflict'. *British Journal of Industrial Relations*. Vol. XVIII.

Streeck, W. 1985. 'Industrial Relations and Industrial Change in the Motor Industry'. Public Lecture, School of Industrial and Business Studies, Industrial Relations Research Unit, University of Warwick.

Streeck, W. (ed.). 1985. 'Industrial Relations and Technical Change in the

British, Italian and German Automobile Industry'. Wissenschaftszentrum Berlin: Discussion Paper IIM/LMP 85–5.

Williamson, O.E. 1979. 'Transaction Cost Economics: The Governance of Contractual Relations'. *Journal of Law and Economics*. Vol. 22.

4

New Technology and Social Networks at the Local and Regional Level

Reinhard Lund and Jørgen Rasmussen

Introduction

The promotion of new technology as part of local restructuring is to an increasing extent the result of the activities of a large number of social actors. Previously managers were the sole initiators of technological change. Recently it has been emphasized that innovations and their diffusion are dependent upon interactions between producers and users (Lundvall, 1985). Furthermore, nation-states have become interested in new technology as part of national policies towards strengthening their economies. Lately, these policies have become supplemented by initiatives by local and regional governments under the pressure of unemployment. It has also been observed that employers not only rely upon themselves but also use their business and employers' associations to develop new strategies of product development, production, and marketing, and arrange conferences for the diffusion of such strategies. In this process of restructuring, trade unions come under pressure to change from wage bargaining agents to political actors working out and negotiating technology agreements and industrial policies. So industrial relations are becoming connected to industrial restructuring (Streeck, 1986: 23). A better conceptual understanding of this new feature of modern society, and especially of the role of trade unions in it, is required.

This chapter focuses upon the promotion of new technology by local and regional governments, bringing together the various actors mentioned above. In Denmark this is done through industrial development councils. A theoretical concept which responds to this is the *interorganizational network*, which will be discussed with regard to the actors and their relations; the network environment; and the dynamics of the network in relation to this environment. We provide an illustration of the conceptual apparatus using the behaviour of trade unions in the local development councils, and within the local network in general, as a case study. Finally, the behaviour of trade unions is evaluated against the background of a strategic model.

This chapter is based on a study of local and regional business development in the North Jutland district in Denmark. The specific aim of the study is to understand the role of local and regional networks in business development. It focuses on three local communities with a population of between 10,000–20,000 inhabitants and, on the regional level, approximately 500,000 inhabitants. The method comprises 3 rounds of interviews with a group of 75 respondents. The study should be concluded in the spring of 1988.

The Network Concept

Defining the Population
As a point of departure we use a definition by Aldrich (1979: 281): 'an interorganizational network consists of all organizations linked by a specified type of relation, and is constructed by finding the ties between all organizations in a population.' This definition focuses upon the population of organizations, the links or relations between them, and the different types of such relations. We add to this the environment and the dynamics of the network. The present section concentrates upon the population of organizations.

In Aldrich's definition the population is taken as given, but actually the delimitation of a network is a problem. In classical sociometric analyses a class of children or the inhabitants of a local community have been the natural objects of network analysis at the individual level. In recent decades specific public programmes, such as labour market training and placement (O'Toole, 1983), have become the topic of investigation, with the network population consisting of those organizations taking part in such programmes. A third way of defining the study population was adopted in a study of the economic development of an American region (Anderson, 1976: 311) when the universe of organizations was constructed by having key informants nominate significant organizations.

In our investigation of new technology and industrial networks two methods are combined. The industrial development officer of three local communes (municipal and rural areas) was taken as a reference point and his contacts traced. Furthermore, one or two major local projects aimed at stimulating the application of new technology were selected, and the actors taking part in these projects were mapped. In this way the usual industrial relations actors were brought into the picture on the basis of their role in local restructuring.

The Contents of Relations
The delimitation of a network is impossible without reference to the content of the relations between the actors. From the point of view of the industrial actors, networks include those parties with which they exchange money, goods and services of an economic kind and/or knowledge and information and/or influence.

For example, the plans for a technology project depended on regional government money, on exchange of information between the firms participating in the project, and on the local government and its industrial development committee providing status to the project.

When using the concept of network in the area of local restructuring and technology transfer, it should be stressed that such a network is characterized by multiple relations in two ways. First, the links between the same two actors contain more than one type of content, and secondly, different links in a network carry different contents. For example, relations between two firms may start with one supplying machinery to the other, whereas later they may exchange information about marketing and supply conditions which are of interest to each of them in connection with their product development. Still later, the firms may together influence the local vocational training school in its planning of technical courses, and this will affect their future recruitment possibilities and their technological development potential.

Environments
The concept of interorganizational network implies an extension of the research from the individual organization to its environment. But networks also have environments.

The environment can be conceptualized as both an objective phenomenon and as perceived and enacted by the network actors (cf. Pfeffer and Salancik, 1978). In this chapter we take the second position. Among environmental factors having an impact upon the actors, we include legislation on local governmental structure and support schemes for specific industrial activities. Another factor is the community culture regarding personal relationships and the amount of co-operation and conflict. Here both community activities in general and specific areas of politics, labour market issues etc. are considered by the actors. A third factor is the local economy where important dimensions are the level of unemployment in the area, the income level of the inhabitants, and the expected level of profitability. If problems arise in any of these aspects, one or more groups of network actors will press for changes so as to improve the unsatisfactory situation. Finally, new technology is felt to have an impact upon the competitiveness of firms and on the qualifications of the workforce, and also upon the interrelationships between firms, research and development institutions, local governments and their actors.

Besides perceived pressure from the environment, the industrial network may also be influenced by outside actors entering the network and creating new contacts. They may do so because of favourable opportunities produced by specific support programmes for entrants, such as the economic support it is possible for firms to get from the Regional Fund of the EC.

Between the network and its environment various exchanges take place. It is exactly because of their expected effectiveness that some national industrial policy programmes are increasingly aimed at strengthening local networks. Such industrial policies are increasingly directed towards training and the creation of new organizational structures in the labour market and in industrial relations.

Interorganizational Dynamics
In studying the dynamics of industrial networks, it is necessary to distinguish between factors in the environment and in the network itself. The latter can further be subdivided into structural factors and characteristics of single actors and their interrelationships.

One network investigated was activated by outside building construction firms and financiers who took advantage of profitable local conditions offered by local governments and others. Another stimulus for local network building was an initiative by the regional government to support local projects promoting new technology. To some extent environmental factors have also been barriers against local development. One example is a prohibition against using municipal funds for economic support of single firm projects.

Within industrial networks it has been found that the establishment of an industrial development council or committee in co-operation with local government has provided a focus for industrial development discussions including the introduction of new technology. Furthermore small councils have turned out to be more effective than larger ones, since committees with more than, say, ten members divided into a small executive and a larger group which was dissatisfied with the information provided and with its largely passive role. It has also been important for the councils to have their own full-time industrial development officer rather than a part-time officer working also for one or two councils in neighbouring communes.

The relations between industrial actors have a dynamic potential. In one of the communes investigated it was stressed by our informants that public support for the construction of physical plant attracted electronics firms to the community, and their success attracted others. This also activated local government to expand housing, schools etc. to the benefit of local building firms and commerce. Furthermore, commercial relations expanded into information and knowledge processes, and mutual trust between public and private actors developed. Again this trust was an asset upon which further initiatives could be based.

The characteristics of the individual actors and their roles in the network structure contain dynamic factors of their own. In the investigation actors were found who regard it as their specific responsibility to develop local industry, to create jobs and growth of production, and extend the use of new technology. These actors were based in local government and in the local development council. In one case they included the mayor, the executive officer of the local administration, the local development officer, the chairman of the local development council, and a politician of the opposition party. Seen as a group, these actors not only control all information about local development but also take initiatives of their own. They function as gatekeepers, liaison actors, and 'bridges'. Gatekeepers control the flow of information into their organizations; liaison actors stand between two or more organizations and act as linkages between them; and 'bridges' transmit transactions in their capacity as members of two or more organizations.

Summing up, the main components of the conceptual apparatus are:

1. delimitation of the network population
   spatially defined group
   a reference point
   joint programmes
2. contents of network relations
   goods and services, financial transactions
   information and knowledge
   influence
3. dimensions of the network environment
   legislation
   culture and politics
   economic situation
   technology
4. interorganizational dynamics
   environmental factors
   network structure
   positive effects
   mutual trust
   actors' characteristics
   dominant actors

It is within this frame of reference that it is possible to analyse the trade unions in a Danish region. A short description of the Danish union structure may be necessary. The Danish trade union movement comprises a number of unions established according to craft principles and four industrial unions. All of these are affiliated with the Danish Federation of Trade Unions (LO). The largest unions are the general union for non-craft workers, the commercial and office employees' union and the metal workers' union. Each has locals at the municipal or inter-municipal level. Below this level are found union clubs and shop stewards at the larger companies. Both at the municipal and county level the unions co-operate across craft lines in local and regional joint organizations which co-ordinate activities, for example on industrial development.

The New Challenge for Trade Unions

The increasing manifestations of deliberate national, regional, and local industrial policies giving priority to new technology must be seen as a new challenge for trade unions. Collective bargaining on wages and hours, and administration of agreements were the classical trade union functions. During the 1960s and 1970s the introduction of new technology accelerated, and in 1981 the first national technology agreement was negotiated between the Danish Employers' Confederation and the Danish Federation of Trade Unions. This agreement was renegotiated in 1986. Both parties accept that the 'use and development of new technology . . . can improve competitiveness, employment, the work environment, and job satisfaction.' The agreements regulate information disclosure of technical, economic, personnel, training and work environment consequences regarding major technological changes (cf. Lund, 1983: 148). At national and local union level, they have been followed by the establishment of technology committees and technology officers in the larger unions.

The new industrial promotion policies are an extension of previous technology initiatives by individual employers, and go beyond the scope of the technology agreements. This raises new problems for the trade unions which have to concern themselves with the internal restructuring of individual firms, but also with changes of industry structure, employment opportunities with regard to skills and sex, impacts upon regional and local housing patterns, demand for social institutions, etc. Industrial and political issues have become explicitly interwoven. The new industrial policies have thus dealt another blow to the traditional division of labour in Denmark between unions fighting on industrial relations matters and the Social Democratic Party taking care of political matters – a division of labour which had already become shaky under the impact, first, of incomes policy and, then, of the 'municipalization' of the Danish welfare state (Tonboe, 1986). These changes have activated the unions at regional and local level, but have also caused problems in their relations with the Social Democratic Party. Unions, for example, often got their own leaders elected as members of local government. Disagreements on specific issues will be discussed in the next section.

Trade Unions in Local and Regional Networks

We shall now analyse the participation of trade unions in more detail at the local and regional level. The following questions will be addressed:

(a)   to what extent do unions take part in the industrial policy network?

(b)   what sort of relations do unions have with the other actors and the   environment?

(c)   what part do unions play in the dynamics of the network?

(a) Network Participation of Unions

At regional level the trade union movement is represented on the regional industrial development council and its executive committee. When the local council in North Jutland took the initiative in a discussion of regional industrial development in 1985, the trade unions were also invited. In addition, the unions are represented on the local councils' employment committee, and on the regional labour market board which is a state authority. The representatives are elected by the regional association of trade unions which is structurally connected to the Danish Federation of Trade Unions. These positions bring the union movement into contact with county politicians, county administrators, leading employers, top officials from the labour market administration and the vocational training schools of the region, and industrial development officers of the local communities.

At local level, the dominant actors have been identified in a previous section. The trade union movement was represented on the local industrial development councils (IDC) of the three communities studied. One of the IDCs has an executive committee where the trade unions were also represented. IDC membership brings the unions into formal contact once a month with local politicians and administrators, employers, and the industrial development officer. The strong bond between the unions and the Social Democratic Party gives unions access to preparatory meetings of the Social Democratic members of the local government body. In one community the union representative on the IDC was himself a Social Democratic member of the local government; in another case, he had been an unsuccessful candidate in the local elections. The trade union movement also plays a leading role in some of the organizations with which the dominant local actors have contact from time to time. These include the vocational training schools and the labour market board.

In one of five development projects which we have studied in detail, the union movement had an initiating role in the provision of physical plant facilities for small-scale manufacturing and handicrafts. In the others they played no active role. Usually the union movement is represented in the local networks by leaders of the largest union, the general union for non-craft workers (SID), which has more local offices than the other unions in the rural areas we have studied.

(b) Network Relations and the Environment

Whether unions participate in IDC meetings or have contacts with the industrial development officer, the content of the transactions usually consists of an exchange of information, knowledge and influence, but seldom of economic goods.

Union representatives seek to improve the employment situation and reduce unemployment in local communities. They follow a consensus strategy helped by the initiatives of other actors whose suggestions on industrial policy are in line with the unions' interests. Conflicts are avoided by all parties because it is felt necessary to co-operate given the present adverse economic situation. In cases where outside actors have decision-making power, the union representatives also feel it unwise to show disagreement. In one of our interviews union representatives, when asked about possible conflicts, replied that they felt the researchers were trying to invent conflict situations.

The unions' contributions to the exchange of information and knowledge with the other actors are based upon information from their shop stewards who are confronted with industrial restructuring and new technology; upon relations with their national headquarters and the secretariat of the labour movement's trade council; and a multitude of other channels. Unions are aware of the difficulties of being up-to-date on industrial policy issues. The industrial development officer may use the union as a bridge between their unemployed members and the vocational training schools when a firm asks for workers with special qualifications.

In cases where disagreement cannot be avoided, the unions feel at a disadvantage because they are a minority on the IDCs. Our study showed that unions had to give in on local housing policies and on the location of manufacturing firms. The unions also meet opposition from state legislation and from liberal politicians and employers at local level when they propose increased involvement by local government in business projects.

Union influence at local level is based upon political contacts and the union network, plus formal representation in many corporate bodies. Demarcation disputes among the unions, and disagreements between unions and the Social Democrats, of course reduce the unions' potential for influence. In the region investigated tensions were found only in cases where the Social Democrats were more concerned about protection of the environment than (some) trade unions. The difficulties which such dilemmas raise for union leaders who are also politically active seem to be resolved in a pragmatic way.

One association of trade unions contributed to the financing of a regional development council in 1958. They have also been intermediaries both as liaison actors and bridges when firms needed funds from the credit market because of the unions' channels to national employee funds and their ownership of one bank and influence on others. But such arrangements are rare.

(c) Network Dynamics and the Role of Trade Unions
In the communities investigated a number of processes of local industrial development have taken place. The union representatives feel that they take part in discussions on equal terms with other members of the IDCs. In general union representatives have facilitated innovation and development processes by providing information about the labour market and by helping in the supply of qualified workers for new technical tasks.

Industrial development has a positive effect on union membership, and this makes it easier to run trade unions and to take part in more initiatives. Furthermore, the change from rural communities towards manufacturing and service occupations has given the labour parties more votes and more members on local government bodies. This process has a dynamic effect upon further industrial development initiatives of the labour parties, such as projects to train the labour force in new technology and attract new industries.

In one of the three communes the union representatives stated that as the unions acquired new functions which they have not previously performed, it was necessary to change the local union structure, and a local association of trade unions was now being planned. Industrial development activities thus seem to have an impact upon union structure, and inversely a more effective union structure may have consequences for industrial development policies.

Perspectives for Union Strategy

Our analysis of union behaviour in the local and regional promotion of technical and economic innovation can be summarized and evaluated by a strategy model drawn from business economics (figure 4.1). According to this model an actor's strategy is formulated on the basis of a set of goals; of the opportunities, constraints and threats offered by the situation; and of the strengths and weaknesses of the actor's structure and resources.

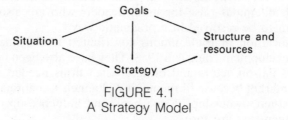

FIGURE 4.1
A Strategy Model

According to one of the leaders of a regional trade union association, it is its main goal with respect to industrial promotion policies to create 'good' jobs which are well paid. To the extent that members are laid off because of new technology, retraining and severance pay must be available. The unions express concern about an unregulated and indiscriminate acceptance of new firms by local communities. Instead they want a priority list. Here they are up against other local actors who stand for more liberal industrial policies. The example underlines the need to see the goals of one actor in relation to those of others. Our study has found that local government bodies have not had discussions on goal setting whereas a start has been made in this respect by the IDCs.

The local situation of the unions has been favourable to the initiation of measures to promote industrial activities. But as already mentioned, the scope of these measures is narrow because of state legislation and a dominating liberal ideology. Yet there are even conservative politicians and employers who are eager for a more active industrial policy as they see the need to compete with neighbouring communities for new business opportunities and jobs. It is also generally felt that the new tehcnology has changed the competitive situation for small and medium sized firms which require a kind of local support which more general national industrial policies cannot offer. The major problem for the unions, then, is to respond when other actors take new initiatives or when new firms or environmental pressures create new challenges.

Our data show that the structure of trade unions at local level is unsuitable for an active strategy towards industrial development and the promotion of new technology. Even though local technology committees are found in some unions, they function only with difficulty. The main impression is that unions have few possibilities to influence industrial development and technology at local level – which is in contrast to the picture one gets from resolutions of national union congresses and from agreements with the employers. Nevertheless, unions have become members of developing local industrial policies networks, and they have strong incentives to try to meet the new challenge.

References

Aldrich, H. 1979. *Organizations and Environments.* Englewood Cliffs, New Jersey: Prentice-Hall.

Anderson, R.C. 1976. 'A Sociometric Approach to the Analysis of Interorganizational Relationships'. *Interorganizational Relations.* Ed. W.M. Evan. Harmondsworth: Penguin, 307–24.

Lund, R. 1983. 'Technological Change and Industrial Relations in Denmark'. *Bulletin of Comparative Labour Relations.* Vol. 12, 139–57.

Lundvall, B.Å. 1985. *Product Innovation and User-Producer Interaction.* Aalborg University Press.

O'Toole, L.J. Jr. 1983. 'Interorganizational Co-operation and the Implementation of Labour Market Training Policies: Sweden and the Federal Republic of Germany'. *Organization Studies.* Vol. 4, no. 2, 129–50.

Pfeffer, J., and G.R. Salancik. 1978. *The External Control of Organizations.* New York: Harper and Row.

Streeck, W. 1985. 'Industrial Relations and Industrial Change in the Motor Industry: An International View'. Public Lecture in European Industrial Relations. Coventry: University of Warwick.

Tonboe, J. Chr. 1986. 'On the Political Importance of Space. The Socio-spatial Relations of Trade Unions'. *Acta Sociologica.* Vol. 29, no. 1. 13–30.

Part II

Innovation Initiatives

5

Management Strategy: Towards New Forms of Regulation?

Serafino Negrelli

Market, Technology and the Organization of the Transitional Company

Contemporary sociological controversies include the nature of the enterprise in post-industrial society. From the beginning of the 1970s the increasingly competitive world market and the development of scientific and technological knowledge rendered obsolete the traditional interpretative models of the enterprise founded upon the concepts of oligopolistic concentration, the dependence of technology upon demand, economies of scale, entry barriers, and optimum size. The criterion underlying the organizational principle that defined the capitalist firm in the period 1930–70 – the capacity of oligopolistic regulation and planning of the market – has also been abandoned and replaced by the concepts of technological convergence and the product cycle. This had previously been used to explain the phenomenon of the horizontal diffusion of innovations during the nineteenth century (Rosenberg, 1976).

Economists have elaborated an evolutionary theory of the enterprise (Nelson and Winter, 1982) which is consistent with more complex and satisfactory models of the continuous interactions between technological progress, the economy, and institutional interests. New computer technologies are primarily standard systems – polyvalent, flexible and adaptable – that allow large companies to cope with instability and greater contextual uncertainty with the same speed and reactive capacity that have always distinguished the strategies of small companies.

Modern organizational theories have developed the notion of flexibility of new production relations either towards the external environment or within the internal organization. Flexibility with respect to the market entails the supply in real time of quantitative and qualitative variations. The Fordist discontinuity between production and sale is replaced by a direct link that attempts to satisfy immediately the demand for personalized or new products. Theoretically, the organization may be subdivided into its two dominant components – the flow of products and the flow of information. The latter is automated and permits the arrangement of an organic and centralized system of information, or rather the requisites and resources for the processes of optimization. It is easy to understand how such a system in turn facilitates the maintenance of a heterogeneous and flexible flow of products. The automated information system can be considered the new 'autocratic' element – the equal of the steam engine that centralized the source of power and of the assembly line that centralized production flow – which through its programs controls the manufacture and movement of products. The tendency to uniformity shown by production systems will facilitate the revision of the classical organizational theories, for example that of Joan Woodward.

Norms, Participation and Power in the Remodelled Enterprise

Alvin Toffler (1985) has outlined the development of the organization of the corporate dinosaur into the new remodelled enterprise. He notes the decline of growth-oriented managers compared to those who confront discontinuity in the environment – undertaking non-linear strategies, possessing a systematic vision of the problems, and open to hitherto unthought-of solutions. The *'ad hoc'* organization supplants that governed by bureaucratic precedent; and the worker must be involved in the co-operative process because, in contrast to the position of Taylor, neither the manager nor the subordinate know in advance the best solution to the problem (Simmons and Mares, 1983).

In the USA this distinctive management behaviour, more pragmatic in its acceptance of hypotheses that can potentially increase the level of economic performance and constituting above all 'a way to run a business', is producing a subterranean reversal in the traditional system of industrial relations. The old adversarial ideology is modernized, the values of the American cultural inheritance are renovated by themes of workers' participation, joint productivity committees are less rare and the 'quality of working life' movement spreads; while at the same time the prejudiced objection to the presence of unions on company boards is reduced. It is useful to take the case of the USA, in contrast to the better known examples of Sweden or West Germany which are traditionally oriented to the constitutional participation of workers' representatives, because in America the post-industrial society is more developed, and collaborative relations at the base co-exist with the neo-liberal policies of the state. Perhaps too much weight has been given in the past to the models propagated by welfare economics. It is forgotten that the American trade union movement derives its ideas not only from the Webbs or Barnett, but also from the institutionalists of the 'Wisconsin School' – Commons or Perlman – for whom trade unionism is principally a social institution to secure control against market competition and technological change.

Today the traditional method of negotiating collective agreements illustrates the inadequacy of pluralist institutions and policies (Cella, 1980). A brief historical comparison between the evolution of the enterprise and of company collective bargaining explains the need to adopt a more diffuse and informal conception of the latter. In the original formulation of the Webbs, collective bargaining constitutes an alternative to individual bargaining; a method of union action to control the conditions of work; an exclusively economic institution using a model based upon the neo-classical theory of the competitive enterprise. A normative process that does not exist in individual, economic bargaining emerges within the activity of collective bargaining. Thus Flanders (1970) makes a distinction between the economic institution of bargaining and the political institution of negotiation, defining collective bargaining as a rule-making process that involves a power relationship between organizations. The model can be related to the characteristics and requirements of oligopolistic enterprises.

In the company which introduces systematic computerization, collective bargaining tends to be extended during the early phase of information and consultation, and this process continues during the later phases of implementation and verification. Continuous negotiation on substantive issues (restructuring, productivity incentives, training, hours and conditions of work) requires, not so much the production of rules, but problem-solving or an agreed reformulation of the issues.

According to the economic–normative perspective that is dominant in the oligopolistic enterprise, systems of industrial relations are regulated by distributive or integrative principles (Walton and McKersie, 1965). The ideal type, according to this normative model, is the Fordist enterprise in which work rules define job autonomy. In the USA the combination of Taylorist techniques and a strong single union in the automobile sector has given birth to an intense activity of job control, resulting in detailed and codified rules in the workplace. In the new computerized company these restrictions become out of date because policies are oriented towards problem-setting in order to cope with the uncertainty of the external and internal environment, adapting to change, and increasing productivity. According to Fox (1974) the problem of industrial relations is to transform the high levels of conflict into high trust relationships which are characterized by shared objects and values, the existence of mutual long-term obligations, a free and open information system, and by the renunciation of short-term benefits for a longer perspective.

The general trend towards the transformation of collective bargaining from market relations to managerial relations (Fox, 1966) is confirmed by the changed attitudes of the managers and workers. In order to move from distributive or integrative to participative principles in industrial relations, it is necessary for the parties to review their relationship, modify their norms of behaviour from the exclusive defence of their own interests, redefine the rules of the game, and therefore to reconstruct the existing power relationships. In the more advanced cases of new technology negotiations, the involvement of the workers facilitates an extension of consultation and the dissemination of *ex ante* information beyond that stated by formal agreements. This trend is assisted by the diffusion amongst management of consultative styles of industrial relations which are more attuned to problem-solving than to the regulation of controversies, in contrast to 'traditionalists' and 'constitutionalists' (Purcell and Sisson, 1983). This style corresponds closely to the notion of the 'radical executive' coined by Toffler for the remodelled company, in which the managers' competences are determined by the changing context.

The Limits of the Transitional Structures

The hypothesis that the regulation of industrial relations by formal rules is being transcended is confirmed by the structure and operational processes of the computerized company. Major difficulties emerge in the transitional period. Those countries with long traditions of industrial relations centralized at company level and with single unions holding a majority representation are at an advantage. Thus during the 1970s in Sweden and West Germany, corporatist alliances at company level were strengthened and hierarchical and horizontal integration facilitated the absorption of the problems of transition, so reducing the trauma. In the USA productivity alliances have been initiated principally in the steel and car sectors, but these pragmatic projects are distinct from the institutional approaches of the above countries and Japan. In the latter, company alliances are diffuse and paternalist, and founded upon security of employment for the permanent employees and personal loyalty.

In Britain, France and Italy it is more difficult to adapt company structures to these new principles because alliances between management and unions are hindered by occupational or ideological union pluralism and by the absence of specific traditions. In Italy one of the major debilitating factors, the discontinuity between national and company levels of industrial relations, results from the unions' historical tradition of concentrating political action at the state level when their economic power is too high or too low (Lange, 1979), thus restricting the opportunities for the modernization of industrial relations. By the second half of the 1970s a polarization between company and national levels of collective bargaining had emerged at the expense of the intermediate, industrial sector. However, centralized agreements supported by the state have tended to limit rather than to expand the central role of company collective bargaining, although its vigour has been emphasized by empirical research. Elsewhere moderate pay claims and the right to information promote a more ordered division of jurisdiction between the levels, and the supervision of negotiations at all levels. Collective bargaining is better structured in companies dominated by participative principles (Della Rocca and Negrelli, 1983). These changes co-exist, however, with other contradictory signs which indicate the inability of certain levels of the company to fulfil their function; and joint management of productivity agreements, restructuring, the introduction of new technology, become problematic.

Innovative Forms of the Regulation of Industrial Relations

The centralization of negotiations within the enterprise favours an autonomous structure, but in Italy the resistance to this trend has produced confusion and fragmentation (Cella and Treu, 1982). Probably the fear of creating a similar situation explains in part the singular absence of any attempt at reform in the national Scotti agreement on labour costs of 22 January 1983 (named after the Minister of Labour at that time). This vacuum has not impeded many companies from studying, innovating and experimenting with new ways of regulating industrial relations. Two indicators are useful for this purpose – the higher level of involvement of the unions in the negotiating process and the increasing institutionalization of procedures and joint committees. The first had developed previously but now embraces the formal conciliation procedures whose absence had restricted union participation in the resolution of industrial and group disputes (Cella and Treu, 1982: 201).

The 'hot autumn' of 1969 and the years immediately following were characterized by bitter struggles and rank-and-file initiatives outside union control. They were succeeded by a more open climate, but this developed without any modification of the formal procedures. The behaviour of the actors changed, unions and workers showed less rigidity in negotiations, managers and company directors avoided confrontation. There was an increased respect for conciliation of disputes which replaced the display of power and capacity for struggle (Negrelli, 1982: 506–8).

The changes in behaviour of the actors and negotiators at the decentralized level also serve to dampen industrial conflict. New methods to institutionalize negotiations and regulate the right to strike have their justification in the greater participation of the union in the social control and economic direction of the company. More structured agreements have been introduced in the public sector, such as that signed in July 1984 with the Minister of Transport, Signorile, which outlines a code of behaviour for companies, rules for union regulation of strikes, and specific guarantee clauses. The private transport companies who belong to the employers' confederation Confindustria are in favour of statutory regulation and have refused to ratify this agreement.

The IRI agreement[1] signed in December 1984 by management and unions is also innovatory. It embodies experimental consultative practices by involving the unions in the process of restructuring the sectors in crisis, and consolidating those undergoing development in order to promote the highest degree of consensus. The creation of authoritative joint committees able to give binding and discretionary judgements on the firm's industrial and labour initiatives contributes to the decisive development of the right to information previously applied by only a few innovative managers. The two functions of negotiation and co-operation require different personal qualities – the first emphasizes aggressive temperaments whilst the second is based upon more accommodating attitudes. The new agreements prefigure institutional arrangements which could reduce the disadvantages possessed by these countries without following the logic of single unionism.

Industrial Relations as Barriers to Technical Change

The difficulty faced by an industrial relations system in moving from a normative to a participative basis varies according to the market, technology and organization of the enterprise, but above all it is related to the union–management negotiating culture. This concept of strategic choice (Kochan, McKersie and Cappelli, 1984) as the intervening variable in relation to the external environment and the particular historical traditions of various countries facilitates a dynamic explanation of social behaviour during a period of change within the company.

A similar, theoretical approach necessitates a reconsideration of political intervention in economic activities, of the relation between economy and society. Some sections of management and unions are still suspicious of the trend towards a post-bargaining regulation of company industrial relations. The former oppose a greater involvement of workers' representatives and maintain that excessive demands from below are the foundation of today's economic difficulties growth. For some unions the reluctance to support p formulae stems either from the fear of signing agreements (increase productivity at the sole expense of the workers, or from antagonism to goals outside traditional union objectives. Both these positions, associated with a conventional, conflictual view of industrial relations, obstruct economic and industrial change. They are more obviously contradictory if the specific strategies of the parties to technological innovation and its relationship to industrial development are considered. The degree of complexity of these strategies depends upon the nature and intensity of such relations (Streeck, 1985a; 1985b): traditional industrial relations structures and procedures either hinder, facilitate or are dissolved by the introduction of new technologies.

An extreme version of this latter process is the *neo-pluralist strategy* characterized by management control, for example that of Fiat (Becchi Collidà and Negrelli, 1986). The metalworking employers' organization Federmeccanica has outlined (1984) the 'manifesto' of a post-union industrial relations system in which individual bargaining supplants collective bargaining, and the technical and ideological gains of Taylorism won in the transition from handicraft to manufacture are maintained against pressures for the recomposition of labour. Their object is 'reduced union demands, which act as inappropriate controls and bureaucratic fetters to the free circulation of creativity'; freedom to hire and fire; a range of negotiated and unilateral methods to control work relations; and company discretion over pay related to output, merit or job. The 'new' industrial relations should be characterized by 'a non-institutionalized union presence' and by 'a balanced struggle in which each estimates their power in relation to the other'. The object is a direct relationship between workers and the enterprise for the regulation of the contract of employment, thus rendering union intervention superfluous or even incompatible with the interests of the skilled workers of the post-industrial society.

A second type of adjustment of industrial relations systems is more complex: *the strategy of concession bargaining* in which the union is incorporated in a subordinate role in the process of change. In the USA the logic in such a strategy is self-evident. Companies possess many exit options from the unionized system: the extension of subcontracting, forms of outsourcing, the closure of factories in unionized regions and their transfer to right-to-work states, the opening of union-free subsidiaries, and other union avoidance practices (Bluestone and Harrison, 1982). Abstention from such options is offered to the unions in order to obtain concessions. The pay system moves to a profit-related basis and the focus of negotiation shifts from the job to the enterprise, to security of employment, and information on investments and company strategy.

This method of confronting change might include complex strategies in which the transitional factors are the object of democratic control and agreement. The best known examples are *integrative strategies* where management makes limited concessions in exchange for worker support for company goals. Agreements that redistribute resources within a given decision-making process provide reciprocal but asymmetrical advantages to the parties (Negrelli and Treu, 1985: 33). Company coalitions constructed upon centralized, hierarchical relations between the actors and bargaining levels are examples of this trend. This solution, typical of central European countries and in a paternalist version in the case of Japan, is spreading throughout the USA as a result of a management initiative to redesign industrial relations (Kochan, McKersie and Cappelli, 1984: 20), contradicting the traditional view that management reacts only to trade union pressure. They are reminiscent of the policies of Italian state enterprises where management has always adopted a more positive policy than that in private companies.

Participative strategies are distinguished from integrative strategies in that they create new structures for the resolution of specific questions such as technological innovation. This approach is typical of the industrial relations patterns of Swedish companies where more flexible organization and enriched jobs result from the redesign of work as an integral element in the new technology of production. But these co-operative solutions are also evident in Italy following the IRI agreement (Negrelli, 1985), and in other countries. These agreements embrace managerial relations: *ex ante* social control of the processes of company change (investment, technology, restructuring, decentralized production and sub-contracting); productivity agreements and jointly formulated criteria for the choice of incentives; concerted action on medium- and long-term labour utilization with particular attention to the development of a career structure; and devolution of bargaining within a structure of centralized controls.

The Impact of Technological Change on Company Industrial Relations

What are the factors that compel companies and industrial relations actors to adopt one of these four strategies when faced by technological change? The proactive or reactive role of management or unions? The theory of managerial initiative as the principal determinant of the structure and dimensions of collective bargaining, such as that of Clegg (1976) or as developed by successive authors (Kochan, McKersie and Cappelli, 1984) is only a partial explanation which ignores the strategies of the other industrial relations actors and the structural changes of the 1970s.

Turner *et al.* (1967) posed analogous questions in the 1960s with respect to the automobile sector. They noted that in spite of the considerable similarity of the mass production technology used for vehicle manufacture in various countries, there was no corresponding uniformity in industrial relations or in the pattern of conflict. It was natural at that time to explain these differences in union power between the various car companies by the origins and traditions of national industrial relations systems and the different customs entrenched at the workplace. The Fordist assembly line embodied the features of mass production that were easily exported from the USA to countries such as Italy and France because of the absence of the union power at the workplace, in contrast to West Germany and Britain whose craft traditions limited penetration. Thus workers in various countries subjected to similar technologies have been unionized in distinctive ways, reflecting either the effective resistance of the work community (Piore and Sabel, 1984) as in Germany or Japan; or management success in imposing bureaucratic relationships at the workplace as in France or Italy.

This analysis of the role of technological innovation in the development of industrial relations systems is consistent with the classical theories of American sociologists such as Blauner, Chinoy, Walker and Guest who first attempted to explain the different degrees of workers' alienation; and of European sociologists such as Friedmann, Touraine and Woodward who related work organization to production systems. However, these are now inadequate explanations of the industrial relations policies of the modern company because accumulated, specific traditions interact with the new computer technologies, which are more complex and flexible than the traditional Fordist assembly line. The origin of the choice of one of the four strategies outlined above is to be found in this interaction. If a historical tradition of collaboration between management and union at the various levels of the company does not exist, then it can prove difficult to adopt an integrative or participative scheme. Elsewhere, the greater costs in time, resources and research required to modernize an industrial relations system limit the use of unilateral or regressive strategies and non-optimal long-term solutions.

The concept of historical tradition in industrial relations is still difficult to define. Managerial behaviour at the macro level is only one element, other variables being the role of the union, legislation, the adaptability of the workers to more flexible socio-technical systems, and the values and ideologies of the different social actors (Bendix, 1956; Kochan, McKersie and Cappelli, 1984). The hypothesis that particular company traditions influence industrial relations strategies is, however, confirmed in everyday language. For example a representative of Federchimica (the Italian chemical employers' association) in commenting upon the speed of the renewal of the national contract, admitted that

> we have had help from the unions when we have had to face the difficult period of restructuring and there is no purpose now in slamming the door in the face of these same people. Each organization has its history and traditions and therefore just as we have concluded a company agreement so we will do the same with the national contract (*Il Sole 24 Ore,* 25 March 1986).

Today it is often more difficult to correlate the features of craft or mass production with the strategies of the various social actors. In the USA, besides the unilateral or concession bargaining strategies, company coalitions are growing. Contradictions exist even in France and Britain where similar company methods of controlling technological progress and robotics in the factory are associated with radically distinct governments (Grunberg, 1984; Winch *et al.*, 1985). In Italy there is a growing divergence between the neo-pluralist model of Federmeccanica and the more participative one introduced by the IRI agreement and that found in the state sector. This division also exists between private companies – for example Fiat and Pirelli – and even within the public sector. Moreover, in many small companies, where traditionally union power has been weak, a new model of concession bargaining is emerging based upon more flexible work roles.

The historical specificity of industrial relations within a company is becoming more important than the type of ownership, the productive sector or national tradition. Technological change and the process of its introduction must be evaluated in the context of this history, and the result of the impact is embodied in the four types of industrial relations strategy delineated above. Is it possible to forecast future possible outcomes or to distinguish at least one or two of the many alternative responses to the acceleration of technological progress?

In the modern enterprise increased productivity and job security are becoming incompatible in the absence of a developed system of industrial citizenship. A more rapid rate of technological change could polarize the industrial relations strategies of companies between a neo-pluralist model in which the unions are weakened, and a participative model marked by single, powerful unions and centralized control of negotiations. Computer technologies are ambiguous in their consequences (Gallino, 1983): they have the capacity to improve the quality of working life and to contribute to a democratic company structure, but they are also capable of facilitating a more refined process of Taylorism in the division of labour. In the long term integrative and concession bargaining models will prove contingent and inadequate solutions, and then the post-bargaining regulation of company industrial relations can only be either unilateral or participative.

Note

This chapter was translated by Paul Smith, who would like to thank Richard Hyman and Serafino Negrelli for their comments on the translation.
 1. IRI is the main group of state-controlled enterprises, covering 400,000 employees in a range of industries. The December 1984 agreement with the three main union confederations was a radical initiative to encourage worker participation.

References

Becchi Collidà, A., and S. Negrelli. 1986. *La transizione nell'industria e nelle relazioni industriali: l'auto e il caso Fiat.* Milano: Angeli.

Bendix, R. 1956. *Work and Authority in Industry.* New York: Wiley.

Bluestone, B., and B. Harrison. 1982. *The Deindustrialization of America.* New York: Basic Books.

Cella, G.P. 1980. 'Per una critica del pluralismo'. *La contesa industriale.* Ed. H.A. Clegg, A. Flanders, and A. Fox. Roma: Edizioni Lavoro.

Cella, G.P., and T. Treu. 1982. 'La contrattazione collettiva'. *Relazioni Industriali.* Ed. G.P. Cella, and T. Treu. Bologna: Il Mulino, 159–214.

Clegg, H.A. 1976. *Trade Unionism under Collective Bargaining.* Oxford: Basil Blackwell.

Della Rocca, G., and S. Negrelli. 1983. 'Diritti di informazione ed evoluzione della contrattazione aziendale (1969–1981)'. *Giornale di Diritto del Lavoro e di Relazioni Industriali.* Vol. 5, 549–79.

Federmeccanica. 1984. *Sindacati e no.* Milano: Il Sole 24 Ore.

Flanders, A. 1970. 'Collective Bargaining: A Theoretical Analysis'. *Management and Unions: The Theory and Reform of Industrial Relations.* London: Faber, 213–40.

Fox, A. 1966. *Industrial Sociology and Industrial Relations.* London: HMSO.

Fox, A. 1974. *Beyond Contract: Work, Power and Trust Relations.* London: Faber.

Gallino, L. 1983. *Informatica e qualità del lavoro.* Torino: Einaudi.

Grunberg, L. 1984. 'Workplace Relations in the Economic Crisis: A Comparison of a British and French Automobile Plant'. Paper presented at the ASA Meeting, San Antonio.

Kochan, T.A., R.B. McKersie, and P. Cappelli. 1984. 'Strategic Choice and Industrial Relations Theory'. *Industrial Relations.* Vol. 23, Winter, 16–39.

Lange, P. 1979. 'Sindacati, partiti, Stato e liberal-corporativismo'. *Il Mulino.* Vol. 23, 943–72.

Negrelli, S. 1982. 'La Pirelli dopo l'autunno caldo'. *Giornale di Diritto del Lavoro e di Relazioni Industriali.* Vol. 4, 435–516.

Negrelli, S. 1985. 'Le relazioni industriali in un'azienda post-industriale'. *Industria e Sindacato.* Vol. 27, December, 13–19.

Negrelli, S., and T. Treu. 1985. 'I diritti di informazione nel processo di democratizzazione delle decisioni di impresa'. *I diritti di informazione nell'impresa.* Ed. T. Treu, and S. Negrelli. Bologna: Il Mulino, 9–87.

Nelson, R.R., and S.G. Winter, 1982. *An Evolutionary Theory of Economic Change.* Cambridge, Mass.: Belknap.

Piore, M.J., and C.F. Sabel. 1984. *The Second Industrial Divide.* New York: Basic Books.

Purcell, J., and K. Sisson. 1983. 'Strategies and Practices in the Management of Industrial Relations'. *Industrial Relations in Britain.* Ed. G.S. Bain. Oxford: Basil Blackwell.

Rosenberg, N. 1976. *Perspectives on Technology.* Cambridge: Cambridge University Press.

Simmons, J., and W. Mares. 1983. *Working Together.* New York: Knopf.

Streeck, W. 1985a. 'Introduction'. *Industrial Relations and Technical Change in the British, Italian and German Automobile Industries.* Ed. W. Streeck. Wissenschaftszentrum: Berlin. Discussion Paper IIM/LMP 85–5.

Streeck, W. 1985b. *Industrial Relations and Industrial Change in the Motor Industry: an International View.* Public Lecture. University of Warwick. 23 October.

Toffler, A. 1985. *The Adaptive Corporation.* New York: McGraw-Hill.

Turner, H.A., G. Clack, and G. Roberts. 1967. *Labour Relations in the Motor Industry.* London: Allen & Unwin.

Walton, R.E., and R.B. McKersie. 1965. *A Behavioral Theory of Labor Negotiations.* New York: McGraw-Hill.

Winch, G., A. Francis, M. Snell, and P. Willman. 1985. 'Industrial Relations and Technological Change in the British Motor Industry – The Case of the BL Metro'. *Industrial Relations and Technical Change in the British, Italian and German Automobile Industry.* Ed. W. Streeck. Wissenschaftszentrum: Berlin. Discussion Paper IIM/LMP 85–5.

6

Between Fordism and Flexibility?
The US Car Industry

Stephen Wood

Computerized technology will herald the end of Fordism. This view is shared by many commentators, as flexible and more automated systems of manufacture appear more feasible. The car industry, the birthplace of Fordism, has figured strongly in such conceptions, as the technological and product markets change, and the relative success of Japanese car firms suggests that in the 1980s we entered a period of rapid innovation. In this chapter I will concentrate particularly on some developments in the US car industry, which has seen major new initiatives in labour relations and work organization through, for example, employee involvement and 'quality of working life' programmes.

It remains an open question whether the automobile case will 'serve as a pattern setter' in American industrial relations (Kochan *et al.*, 1986: 242), as it did in the post-war period; and indeed proponents of various theories of work reorganization have all used the car industry as an example to support their case. These range from Shaiken's (1985) analysis of the new technology largely in terms of the Bravermanesque control perspective, to Guest's (1979, 1983) and Walton's (1985) portrayal of employee participation as a reflection of managements' conversion to a commitment-type organization, in response to workers' rising educational standards and aspirations for industrial democracy. The problems of the car industry have also figured strongly in discussions of the crisis of mass production in Piore and Sabel's (1984) flexible specialization thesis (see Hyman, chapter 2 above) which emphasizes the potential of the new technologies for reskilling leading to more rewarding work. Piore and Sabel's orientation towards the demise of Fordism leads to a temptation to contrast their thesis with labour process theory, with its emphasis on the dominance of Taylorism

pitalism and management's labour control problem. Both the
process and flexible specialization arguments are important
they question the neo-human relations treatment and romantic-
izatiᴏɴ of the 'new' forms of work organization. The danger, however,
is that debate is based on stereotypes of the opposing theories and
becomes polarized or simply reduced to definitional questions. It may
be more useful to develop the many strands of argument which lie
between the two extremes.

First, we should acknowledge that new forms of work organization
co-exist with Taylorist practices. In some circumstances the alternatives
may amount to genuine reversals of the basic features of scientific
management; in others they may only represent modifications; or even
simply a gloss on the underlying Taylorist approach which labour
process theory assumes to be universally present. Taylorism should be
treated as an historical force which can take on different forms in
different contexts. This allows for different adaptations, modifications
and developments. At the heart of the problem of Taylorism is the
vital assumption that conception can be reduced to a matter of science.
There are limits to this possibility and for this reason the divorce of
conception from execution will never be total, its extent will be a
matter of degree (Manwaring and Wood, 1985). Job redesign programmes should thus be treated multi-dimensionally and they may
reverse certain features of Taylorist practice whilst reinforcing others,
a point which is perhaps underplayed by Armstrong (chapter 8 below)
and others who defend the control model.[1]

Second, we should acknowledge that, as the flexible specialization
argument has come to the fore, it has become increasingly clear that
Piore and Sabel's thesis is based on an extreme model, and that their
vision of a future, what Hyman calls the new panacea, dominates their
analysis of the present. Their thesis exaggerates the flexible capacity
of new technologies; and there is an underlying, inadequately explored
assumption that increasing fragmentation and complexity in the product
market 'demands' flexible technology, which in turn 'demands' the
flexible worker, who is seen as a kind of multiskilled craftsman. Other
commentators (Altshuler *et al.*, 1984; Friedman, 1983; Katz and Sabel,
1985; Streeck and Hoff, 1983; Tolliday and Zeitlin, 1987) have
acknowledged that the precise outcome of the new flexible technology
depends on the corporate strategies adopted by the leading firms, as
well as national differences in industrial relations and training systems.
Their work also points to the relatively small number of examples of
new initiatives; the ambiguities in the way in which managements
respond to the new technology; and the fact that some managements
(Fiat, for example) have attempted to develop more involved workforces
without a co-operative strategy towards the unions. Much writing on
the flexibility question retains, nonetheless, an implication *à la* Piore

and Sabel that in so far as firms do decide to use the potential flexibility of the technology, they will embrace the full flexible specialization model or something approaching it. Positions between Fordism and flexibility will then reflect either a transitional period or a hesitancy on behalf of managements in taking on board a completely new labour strategy.

But ambiguities in managements' labour policies may reflect genuine doubts and uncertainties as well as the multi-dimensional nature of corporate strategies. The consequences of the technological changes may not be clear-cut, and technologies should not be treated as having precise effects which are known to the actors involved. Part of the argument of this chapter is that precisely because of this managements may, as the rate of technological change rises, and as they aim to introduce model and style changes ever more rapidly, increase the involvement of workers. The new developments in work organization do reflect an emerging, more automated era of mass production which demands new political and social institutions, at least within the factory. They are not simply extensions of Taylorism at a higher level of automation, as labour process theory implies. But, whilst the quality of work life experiments 'have less to do with concerns for the workers' contentedness than with the need to reduce the rigidity of existing assembly procedures' (Sabel, 1982: 213), they need not imply an abandonment of mass production or scientific management methods, as the extreme flexible specialization argument suggests.

Worker Participation in the US Car Industry

The Worker Participation Scheme
An industrial relations strategy must address the various tensions which result from environmental pressures and the product strategies adopted by the US car companies. Examples include the problem of justifying concessions when large investments are being made abroad; of demanding relative wage cuts whilst asking workers for more effort, care and responsibility; the tensions involved in developing policies which link rewards more directly to plant and individual performance within the context of a long-established and highly valued (by the union) pattern bargaining arrangement; the ambiguities and fears created by changes in sourcing patterns; the establishment of joint ventures and the opening of new plants in the largely non-union southern states of the US; the dilemma within training policy, between concentrating efforts on the active workforce or retraining those who are, or are about to be, laid off. Of increasing importance is the tension (inherent in the technology) which issues from the fact that the development of new projects may begin years before the first car

is produced, yet their final conception depends on a debugging and learning process which cannot predate the start of production. Perhaps above all else, there is the tension created by the simultaneous pursuit by managements of both job reductions and productivity improvements from the remaining workforce. Managements' labour strategy has to be directed to such issues – as well as the more specific labour problems such as strikes, working practices, and relative compensation rates.

The main employee participation schemes in the industry are Ford's Employee Involvement Program and General Motors' (GM) Quality of Work Life (QWL) Program. Chrysler's Product Quality Improvement Program is more limited. All three are concerned to improve the overall commitment of all workers to quality and getting the job right first time, whilst simultaneously increasing workers' satisfaction through that enhanced involvement. Unlike the other two programmes, Chrysler's is largely confined to improving product quality and it is thus explicitly not a 'quality of work life' programme – it consists of plant quality improvement committees and a limited number of quality circles, and works mainly through exhortation. This chapter will concentrate on the two more developed schemes of worker participation.

The schemes have a number of dimensions:

1. They are company-wide programmes and are an important part of corporate strategy. As such they can serve a number of functions and allow for variety in their implementation.
2. They are intended to be distinct from collective bargaining which continues to deal with all those matters traditionally the domain of the labour contract.
3. They are run jointly by management and unions, and entail a hierarchy and variety of joint committees and activities.
4. Their prime orientation is towards shopfloor change – particularly in the style of supervision and extent of cost and quality consciousness. This is principally achieved through the formation of problem-solving groups or quality circles.

Although the worker participation schemes are not portrayed as a substitute for bargaining – and the union tries to ensure that contractual matters are not taken up in participation forums – managements have increasingly begun to see them as a way of addressing their industrial relations problems more directly.[2] First, they see them as a means of changing the issues which fuel the bargaining and its underlying spirit. Second, as noted by Katz (1985) and confirmed in my interviews, managements have viewed the schemes as ways of reducing reliance on the contract and the grievance procedure. Third, worker participation has increasingly been used as an alternative, or supplement, to local concession bargaining, particularly over working practices. Fourth, the success of certain initiatives in worker participation, reinforced by the Japanese example, has become a model for the design of labour relations in new or modernized plants.

Attempts in the 1980s to reduce the centrality of collective bargaining have been further encouraged by provisions of the various national bargains at Ford and GM, such as profit-sharing. The commitment of both parties to worker participation has been endorsed and new or revised adjuncts to worker participation have been agreed – for example, alcohol rehabilitation programmes, absenteeism programmes, and plant mutual growth forums which represent plant-wide information sharing and problem-solving institutions (Kochan *et al.*, 1986: 133–4). The job security problem has been addressed through company funded retraining programmes and moratoria on plant closures. In addition the job bank schemes (officially termed Job Opportunity Bank-Security at GM, and the Protected Employment Program at Ford) limit the companies' right to lay off workers except as a result of market changes. If, for example, the production of an engine previously manufactured in a US plant is transferred to a Mexican plant, an equivalent number of jobs to those lost in the US plant will be created in the bank.

It is also important to view the industrial relations changes in the context of earlier attempts to deal with labour problems. Of particular significance is GM's attempt in the 1970s to set up non-union plants in the southern states, which the UAW was able to arrest (Kochan *et al.*, 1986: 60–1, 157–9). In the 1980s GM switched its emphasis to establishing worker participation in the new plants; joint ventures such as the NUMMI plant (with Toyota) in California; and new operations which fall outside the existing labour relations pattern. GM has set up a completely new division, called Saturn, to build a new small car by 1988 or 1989. During the 1984 negotiations, GM undertook to authorize the investment for this project and to recognize the UAW. In late spring of 1985 the site was announced: Tennessee. Whilst this state has a right-to-work law, it is also close to the Nissan plant, and some of the leadership of the UAW saw this fact as a means of developing the union's influence outside its traditional centres of power in the industrial north. As part of the Saturn development a group of workers, with management backing, worked full time in 1984–5 to design the main features of the employment relation and work organization systems. The new plants are being constructed on the team system and so will have no formal worker involvement programmes. Worker participation has thus become intimately involved in a process through which managements are developing new technologies and forms of work organization in what was previously a highly unionized industry.

In assembly plants there has been no significant change in the production system towards a Volvo-type work organization, for

e. Teamworking and quality circles[3] in and of themselves may
:r the essential characteristics of assembly work. As Katz (1985:
δδ–1υ4) shows, even the use of work teams has not led to the
abandonment of assembly line production techniques. The worker
participation schemes involve various types of teams, but they
need not entail complete teamworking, that is, the curtailment of
individualized work stations and job territories (Katz and Sabel, 1985).
But the dominant tendency in the newer plants is for worker
participation and job redesign to be integrated through teamworking.
In GM's Fiero plant the team concept extends to all levels of the
organization. The chairman of the shop committee sits on the
administration team, and there are business and resource teams as well
as operating teams (Kochan *et al.*, 1986: 159–62; 198–9). Despite the
growth of such joint committees in most plants during the 1980s, the
involvement of worker representatives in strategic decisions at plants
such as the Fiero plant remains exceptional, and such examples may
differentiate GM from Ford. (Chrysler does, however, have the UAW
president on its board.) The differences between the schemes at Ford
and GM are nevertheless small, although GM does possibly allow for
more local variation and is more explicitly oriented towards attitudinal
change. The main difference between the two companies lies in the
extent to which GM's corporate power has allowed it greater freedom
to experiment and to follow more varied avenues in both its labour
and corporate strategies. Quality of Work Life, Saturn and other
technological initiatives, joint ventures, both in the US and abroad,
are all part of a highly complex and often ambiguous corporate strategy.

Differences between plants, particularly between the newer and
older ones, are more important from the point of view of worker
participation than any differences between the companies. In one 30-
year old GM plant, for example, the core of the programme is the
quality council which meets every day and comprises 20 representatives
from all the departments. In contrast to this highly specific, task-
oriented approach is the scheme adopted in a plant with a similar
technology and age, where under the programme the entire workforce
went, over a period of nine months, on a three-week orientation course
which included problem-solving and group exercises, as well as sessions
on the overall business situation and the importance of worker
participation. Both these cases can be contrasted with the newer plants
in which the worker participation schemes have a variety of dimensions
and may include a good deal of teamwork, weekly or even daily team
meetings, new forms of grading and payment systems (pay for
knowledge,[4] for example), and a very different style of personal
relations between superiors and subordinates.

The Significance of Worker Participation

The significance of the concept of worker participation is that it may facilitate a variety of adjustments which individual plant managements may feel necessary. Worker participation can achieve a number of outcomes. It can make supervision more relaxed and encourage identification with the plant through, for example, open days and outings, and encourage employees to make suggestions which will cut costs and improve quality. It can enable managements to harness to their advantage the practical or day-to-day knowledge that workers have, and it can facilitate the development of new working arrangements by changing work rules and fostering team work. Many of the proposed changes initiated through the various types of participation groups are concerned primarily with improving quality and reducing costs. For example, they include suggestions for tool changes, improved material handling arrangements, redesigning machines and plant layout, enhancing and guaranteeing the quality of the components purchased, and reducing costs through energy saving schemes and housekeeping arrangements. They range from major proposals which emanate from months of meetings – such as the scheme to improve energy consumption at one Ford plant I visited on which groups of skilled men had worked for seven months, to relatively minor ones – such as the suggestion that the company purchase better quality rags on the grounds that they will last longer and hence unit costs will decrease.[5]

Despite the continuation of the assembly line principle, tasks have been changed in a number of important ways. First, 'half an hour in a meeting is half an hour off the line', as one union representative put it; that is, by giving a worker time off for team meetings or an activity falling under the participation programme the deprivation involved in a day's work is reduced. Second, there are changes in work rules and the potential for increased job rotation and enlargement as the new arrangements allow particular individuals within the same production set-up to have greater mobility between jobs. Third, there are changes in tasks associated with the new technology such as the combination of certain skill groups, a development towards which most car firms have moved. Fourth, much of the new technology has eliminated the most deskilled jobs and also reduced the stress and physical problems of certain jobs. For example, there may be ergonomic changes which reduce the need for excessive bending or stretching. In such cases, labour can, because of changes in production methods and product design, be intensified without any increase in work effort or control. Fifth, there may be changes in decision-making procedures concerning production arrangements, since one of the objectives of worker participation groups is to discuss problems related directly to the job and quality of the product, and to move beyond simply discussing working conditions or complaints about supervisors.

Finally, tasks have been considerably altered by the introduction of modern computer-aided management methods such as statistical quality control. Also, the successful implementation of new technologies and production facilities may depend on workers applying their existing knowledge to the fresh situation, and in most cases their input will be needed in the debugging of technology and the correction of quality faults. Moreover, just as the development of Taylorism demanded certain new 'behavioural traits' (Lazonick, 1983: 112), so does the increasingly automated factory. 'Fitting in' increasingly means having a broader awareness of the way the production systems operate, and usually requires training in statistical process and quality control. The 'diagnostic skills' which are required do not just arrive; they need developing by on-the-job training, particularly through experience of solving problems.

Whilst changes in tasks may reduce the degree of specialization and increase workers' responsibility and autonomy, management's overall control may also increase. For example in the carrier system used at GM's Kansas plant, work groups have considerable control over the pace of work, since the carrier cannot move out of a work station until released by the group; but, even then the group is 'expected to maintain a set pace'. While work groups are able to vary their work pace on individual jobs, 'computers keeping track of the carriers will quickly detect any group that is falling behind, preserving the productivity discipline of traditional lines' (cf. Holusha, 1986: D2). Other features of the managerial strategy also increase management's control and reduce individual autonomy. For example, the joint programme to reduce absenteeism has the significant title of the Absence Control Programme at some GM plants.

There is also an important attitudinal dimension to worker participation and consequently a great deal of symbolism surrounds the programmes. For example, all plants are encouraged to define and disseminate their overall philosophy and objectives. Copies of such statements are posted in strategic places throughout the plant, and in some a plaque in the main entrance displays the plant's objectives. Posters, pens and T-shirts conveying the UAW–GM or Ford's worker participation motif and the importance of quality are readily available. Some plants make use of the notion of a worker participation university from which workers may graduate as, for example, worker participation co-ordinators. The goals of attitudinal restructuring are more than simply fostering increased cost and quality consciousness. Workers are being made aware of plant performance, and the need to compete with producers from anywhere in the world.

Every plant is now encouraged to think in competitive terms. Component plants are especially reminded that they must compete with outside suppliers;[6] and several of the work rule changes both GM and Ford have publicized concern plants threatene outsourcing unless costs were reduced and working practices changed. A participative approach to plant-level industrial relations which allows for diversity in implementation in fact encourages competition between plants and individuals within the same company. Even if a plant does not have a direct rival within the company – as is true for most Ford and Chrysler plants – it is ultimately competing with others for future investment. Comparisons are increasingly made for industrial relations purposes between plants within the same firm and between the performance of different countries (cf. Dohse, 1986). Such processes are part of the attempt to make union representatives understand management's view of the measures they regard as necessary to face competition, and to gain a commitment from their workforces to the changes. Japan lies at the centre of the competitive threat:[7] for example, the GM–UAW joint working party's report on the company's competitive position concentrated almost exclusively on the differential performance of Japanese and US car manufacturers.

The Japanization of the Industry

The characterization of the differences between Japanese management and conventional methods which emerged within the car industry in the early 1980s highlighted the former's superior stock control procedures (Kanban or just-in-time inventory policy); emphasis on quality over production; impressive turn-around times and other indices of production performance; and heavy utilization of workers' know-how. In reacting to the perceived differential performance between Japanese and US car manufacturers, particular attention has been paid by US managers to adopting Japanese management's production methods. Before considering this development further, it is necessary to stress that the Japanization of the US industry is not only a learning process. It involves the opening of Japanese plants in the US (in the case of Nissan and Honda alone or in joint ventures, as in the cases of Mazda, Mitsubishi and Toyota). This has meant Japanese management methods have been introduced directly into the US in the form of non-union plants with lower wage rates and benefit packages, reduced job classifications, fresh working arrangements, and certain forms of teamworking.

The outcome of the negotiations at NUMMI suggest that there will be no major differences between the labour arrangements in Japanese plants even if some of these are unionized. There are three grades for skilled workers and a single one for the unskilled, as compared with the 81 classifications which existed when it was solely a GM plant.

The role of the steward relative to the team leader was initially ambiguous and was contested throughout the plant's first year of operation. But the UAW representatives were able during negotiations in Spring 1985, to establish a representative structure which was independent of the team system in return for their 'acceptance of a large measure of responsibility for quality and efficiency' (UAW–NUMMI – Special Report to Membership June 1985, Highlights of Tentative Agreement). As part of the team concept everyone eats in one canteen and parks in the same car park, and is encouraged to wear the same grey jacket at work. All employees are called team members and the workforce played an important part in perfecting both the training programme and set-up of the production system. During the first few weeks, only one or two cars were sent down the line, while during the following weeks the numbers increased gradually until the plant formally commenced production about six months after start-up.

An important feature of all the Japanese plants is their significantly reduced labour/output ratios. In the Honda plant, for example, the lower employment level reflects not only a high level of automation but also superior production organization and a commitment to reducing labour input. In these plants all employees are required to seek constant self improvement – what is known as 'Kaizen' at NUMMI and Mazda – and selection is highly intensive. All candidates for employment at NUMMI, for example, are required to attend what is called a 'pre-employment workshop', which consists of three days of interviews, job simulations and discussions concerning the firm's objectives and philosophy.

In the US-owned firms, in keeping with the perceived diagnosis of the main causes of the superiority of Japanese enterprises, particular attention has been given to adopting Japanese production methods, much more so than to directly copying Japanese labour relations. To a greater or lesser extent in all plants in the 1980s, computerized systems of managerial control and other Japanese production methods, such as self-inspection and the ethos of 'right-first-time', have been introduced. The extent of the adaption of Kanban varies between plants, but the overall corporate strategies are oriented towards the Kanban philosophy. All plants are increasing the proportion of parts they purchase on this basis. But it is in the newer plants that such practices are most developed and Kanban, statistical process control and other elements of the new production package are central to the overall conception of these plants. NUMMI's management claims to obtain 75 per cent of their parts on a just-in-time basis, and 70 per cent is claimed for the Chrysler Sterling Heights plant. A zero inventory policy was included in the constitution of the Saturn project.

The managers in the US auto firms I interviewed all put great stress on the Japanese example when accounting for the development

of worker participation. The successful adaption of ⌐
production systems is especially seen as depending on
As one industrial relations executive of a US car produce
can achieve a great deal with a workforce in a situation of trust
co-operation, including allowing them to stop the line for quality
problems'. Permission to do so is given to operators at NUMMI and
other new plants. As the plant manager of NUMMI put it, 'it is very
important to have a co-operative relationship between labour and
management . . . based on mutual trust and respect.'[8] Just-in-time
production and employee involvement are then best treated as two
sides of the same coin, the zero slack and defects policy.

But managements are approaching the adoption of Japanese methods
selectively and my interviews with managers in the US uncovered some
recognition of the limits of the usefulness of Japanese-style industrial
relations. Most managers were doubtful of both the desirability of
certain aspects of Japanese practice – the use of dormitories, the
excesses of paternalism, and its extreme patriarchial nature – and the
feasibility of implementing some of their techniques, for example
extending appraisal systems to workers or offering life-time employment
to all (cf. *Harvard Business Review*, 1981: 73). There will nevertheless
be an attempt at Saturn to offer permanent employment to 85 per cent
of the workforce. The attempt, as part of the 1982 agreements at GM
and Ford, to link directly job security to work rule changes under the
pilot employment guarantee schemes was, however, not very successful;
of the eight sites chosen for the experiment, only one site implemented
the scheme in its first three years.

Other more general attempts to guarantee employment, namely
GM's and Ford's 1984 job bank schemes, are seen in some circles as
providing the kind of job security associated with the Japanese life-
time employment. They will, so the argument goes, underpin worker
participation as they guarantee that employees' suggestions and other
aspects of increased involvement will not result in job loss. Certainly
the schemes offer more protection than was previously available to
those whose jobs disappeared because of outsourcing or new technology.
However, the schemes were not seen as offering increased security by
the worker representatives whom I interviewed in plants where there
is a concern about possible future closure, for example, at Tarrytown
where navigation problems on the River Hudson make the plant
permanently vulnerable. Moreover, the schemes do not appear to
reflect a deep new commitment to job security; they did in fact arise
out of negotiation. One of managements' primary bargaining objectives,
to maintain their right to make sourcing decisions, was achieved; and
the representatives of corporate managements I interviewed did not
regard the job bank schemes as a major constraint or as representing
more than a marginal cost compared with existing programmes, such

as the guaranteed income stream. In addition, the schemes do not offer protection against reductions in demand for a particular model arising from increased sales of rival products from the company's overseas operations, joint ventures, or companies outside the main corporate structures of the firm in question. Thus if, for example, increased sales of cars in the Mazda and Ford joint venture are at the expense of existing Ford products, the Ford workers are not protected. Workers on existing GM models will likewise not be protected from the effects of the successful launch of the Saturn car.

The concept of Saturn is itself a reflection of the assumed problems of adapting present social arrangements to meet Japanese competition. The final form of the Saturn organization has not yet emerged, but it is clear that the aim is to go beyond existing experiments, as well as what has materialized at NUMMI. The intention, as stated to me in May 1985 by the working party referred to earlier, is to go beyond both worker participation and Japanese management, and in particular to make the managerial prerogative less clear-cut. From my interviews, it was not entirely clear quite how this will be achieved; for example, the precise role of the industrial engineer still needed clarifying. But workers will be involved in a set of higher level planning committees, such as the manufacturing and strategic advisory committees (Kochan *et al.*, 1986: 201). The role workers or their representatives will have (or have had) in decisions about sourcing, product design, investment and technology still remains an open question. In the joint discussions so far, stress appears to have been placed on productivity and its achievement through Japanese production methods and novel work organization. But the technology is important – an advanced module process is to be adopted – and will be decisive in determining employment levels and skill requirements, as well as the basic parameters of work organization. The package of industrial relations which has emerged displays all the features of the 'new' industrial relations: a gain-sharing scheme, pay for knowledge, single status for unskilled workers, common facilities for management and workers and problem-solving groups, as well as guaranteed employment for core workers. The role of the grievance procedure and committeeman or steward within this system is far from clear and seemed to be an unresolved issue when I interviewed the team. Neither is it totally clarified in the *Memorandum of Agreement between Saturn Corporation and UAW,* dated July 1985. But it does appear that the extreme Japanese practice in which the supervisor has the dual function (cf. Nomura, 1985) of both controlling and appraising workers and acting as their representative, as well as being effectively the centre of the work study system, will not be adopted.

The Flexible Specialization versus World Car Scenario

When writing about the car industry, proponents of the flexible specialization thesis have tended to contrast their scenario of the future with that of the world car, which implies that there will be a new international division of labour. Cars will, according to this perspective, be designed in a small number of firms and in only one or two centres; and the increasing outsourcing and use of newly industrialized countries will result in a division between the type of work done in these countries and the traditional primary countries. Deskilled work will be done in the newly industrialized countries and only the high skilled work will remain in the high wage, developed countries. This implies that there will continue to be job losses in the US and that eventually employment will be only of a highly skilled craft-type.

In contrast, the implication of the flexible specialization model (at least as originally formulated) is that such job losses will be largely averted as more work will spread amongst existing centres of production and the necessity to achieve economies of scale through world sourcing diminishes. In this second model, participatory systems will be especially important as they are a means of breaking down rigidities such as work rules, and an end in themselves since the new type of work organization requires a flexible worker. In contrast, under the world car scenario the worker of the future in the US will be a highly skilled worker, and increasingly involved in specific projects – much as are those currently working in the GM technology centre in Detroit. As such, worker participation seems important only in the management of change, for example in reducing demarcations between skilled trades and in maintaining motivation in plants whilst their workforces are being reduced or other plants are being closed.

The debate about the world car versus flexible specialization has been actively encouraged in the MIT 'Future of the Automobile' study (Altshuler *et al.*; 1984), and its two poles largely mirror the labour process and flexible specialization arguments with which this chapter opened. When it initially developed, such a debate might have been justified. But the issues may be more complex than a straight dichotomy between flexible specialization and world car can capture. In the 1970s international business specialists such as Vernon (1974) were arguing that products like cars had reached a stage of maturity, and that consequently competition would be fought primarily in terms of labour cost. This seemed to explain the success of Japan with its cheaper labour, and it implied that newly developing countries would play an increasing role in world production. The reaction to this kind of argument (as well as to Fröbel *et al.*'s (1980) new international division of labour thesis) has been to point to the importance of product

and not just labour costs or capital's needs for control – in
ecisions (Schoenberger, 1985). Also, attention has been
..ie influence of the political environment – for example,
protectionism in developed countries and political instability in Latin
America (Dankbaar, 1984; Dombois, 1986).

Moreover, particularly following the work of Abernathy (1978), the
concept of the car as a technologically mature product has been
increasingly questioned. Consequently Japan's advantage has come to
be seen as stemming primarily from its development of new production
and design arrangements and its superior handling of the relationship
between product and process innovation, and not principally in its
labour relations. To the flexible specialization theorists, this points to
Japan's superiority in using flexible manufacturing, as they see the
innovations in production methods as pointing to the emergence of a
qualitatively new system of production (cf. Tolliday and Zeitlin, 1986:
20). This is not, however, so self-evident to others. For example Dohse
et al. (1984) interpret the superiority of Japanese production methods
in more conventional Fordist terms: for them, at least as they viewed
it in 1984, Japanese management had perfected certain aspects of
Fordism, although in addition they had developed certain key
innovations like statistical process control and Kanban.

The evidence on Japan is mixed. Japanese firms do appear to
introduce new products more rapidly, their production systems seem
more flexible and the workforce appears to contribute more to the
perfection of these systems (cf. Friedman, 1983). In support of the
Fordist interpretation of Japan we could, following Dohse *et al.*, point
to the way in which quality circles may not violate central principles
of scientific management – providing, that is, they produce suggestions
for improved industrial engineering and rationalization. However, just
as there is a danger of exaggerating the novelty of the Japanese
methods, so there is a danger of minimizing them in the neo-Taylorist
interpretations of Japan. Japanese methods need not be interpreted as
simply a continuity of Taylorism, precisely because they may involve
employees in the conception process. As a response to the impossibility
of reducing conception to the precision of science and determining the
one best production method (Stock, 1980; Manwaring and Wood,
1985), Japanese-style worker participation is perhaps best treated as a
departure from conventional Taylorism, albeit within the scientific
management tradition. It is oriented towards overcoming the limitations
of the Taylorist approach to industrial engineering. The central feature
of participation in Japan at least in the auto industry is, as indeed
Dohse *et al.* suggest, its attempt to combine the specialist and developing
knowledge of the professional engineer with the day-to-day concrete
knowledge of the worker. So, whilst it may, for certain purposes, be
appropriate to place emphasis on their production methods, ultimately

their successful development cannot be separated from both their workforce (cf. Maurice, 1986) and labour re

There is also a risk of underplaying the role of techno competitive advantage. First, there remains the questio to which the new Japanese inspired methods are linked to technological advance. The flexible specialization argument jumps too readily to the assumption that they are linked. In contrast, others have gone too far in denying the role of technology. It is possible, however, to see, as Hayashi (1982) does, the development of worker participation in Japan partly as reflecting adaptation to increasing automation, without assuming the end of mass production. Automatization of production was, so Hayashi (1982: 3) says, 'conditioned on the system of standardization and synchronization . . . which Ford . . . developed in 1914'. Both the costs and problems of advancing automation, even for relatively standardized production, imply a need to reduce downtime and retooling times, as well as a changed relationship between workers and managers and, in particular, industrial engineers. Second, and relatedly, Japanese managements do appear to be strongly oriented towards utilizing and developing advances in technology, particularly in production management. It is hard to accept that the large difference between Japan and the US in both the number of production hours per car and its rate of change can be explained as simply a reflection of their ability to perfect Fordist management control methods, without any reference to technological differences. Observing the leap in robotization in the 1980s, Malsch *et al.* (1984) noted that the Japanese were major pioneers and emphasized that this was not as piecemeal as the introduction of robots had been in the 1970s. As such it represents a major step towards the fully-integrated computerized production system, including the integration of computer-based product development and production planning. It is the increasing computerization of design and production control – the use of on-line computers for controlling the flow of materials, checking faults, and balancing production – and not the automation of given production tasks *per se* which is central to the development of the industry.[9]

When considering flexibility within the context of these new technological developments we must be far more specific than much of the discussion thus far has been, or than I can be in this chapter. The flexible specialization thesis is in danger of assuming that all the new technology is inherently flexible and that this flexible potential is being fully used,[10] and of neglecting the gradual way in which the new technology has been introduced.[11] It is necessary to treat the concept of flexibility multi-dimensionally (Adler, 1985), and as a matter of degree; the emphasis in auto manufacture is on increasing routing, production, volume and programming flexibility within given capital installations and not primarily on product flexibility. Many of the new plants remain relatively dedicated, largely because the differential cost of dedicated versus flexible plants remains high. Furthermore, much of the computer-aided design is concerned to utilize common parts and the degree of standardization of the interior features of cars in particular would appear to be increasing. Modern car design concepts mean that a wider range of products can be conceived within a smaller set of concepts. With modern technology, notably the increasing use of electronics, the complexity of cars has been increasing, although there is also the danger of exaggerating the extent of past standardization. Because of all the model derivatives, engine combinations, colour combinations, option and legal and marketing requirements there is a tremendously high level of parts complexity in all car plants. It is the way in which complexity, as well as quality, is coped with which is of primary importance, and not some overall desire for totally multi-purpose systems which, for example, could switch at low cost from large to small car production.

The uncertainties of the future, as well as of the development process, mean that there is no real desire to achieve a genuine qualitative increase in flexibility. Design concepts remain oriented towards standardization and the technology conceived within the mass production regime. Much of the investment in new technology has, as Tolliday and Zeitlin (1986: 17) say, been 'aimed at improved product quality and process efficiency' and the managers I interviewed in the US all emphasized that the main effects of the new technology are, thus far, considerably improved quality and innovations in production control. Nevertheless there is considerable flexibility built into the design and engineering of the car. Within the overall design concept a particular 'car' – or, as it is known, platform – can cater for a variety of styles, and a given style can be used to serve a number of differing sectors. Furthermore, developments such as fuel injection and turbos also mean that very different performance standards can be achieved with basically the same engine.

The explicit world car strategy exemplified in Ford's Escort and GM's Cavalier in the 1970s may be less prominent, but it could be a mistake to equate a move away from such a strategy with a move away from internationalization, or even away from the world car.[12] The debate about the globalization of the car industry is in fact far from over and, gradual though the change has been, this does not appear to be reversing, although this need not necessarily result in standardized world cars or in a global market. Different companies may pursue different strategies, and individual companies may pursue a variety of strategies. Nevertheless, since the late 1970s, global competition and the global integration of the industry have been developing (Hill: 1983). Internationalization so far involves the co-ordination of production, marketing and design across national boundaries. Many cars are becoming nationless, whilst simultaneously the design concepts facilitate a variety of different models to cater for different market segments and local needs. There is an increasing variety in the country of origin of a car's components, and more use is being made of joint ventures, especially as a way of spreading the costs and risks of design and achieving further economies of scale in engine and component manufacturing. An overall effect of such developments is that both GM and Ford are extending their outsourcing of certain components and presenting to the US market competing cars from all three of their potential sources – their own internal production, their own offshore production, and their joint ventures.

The internationalization issue is compounded by the rise of Kanban which implies a large usage of local suppliers. The global evidence on sourcing is unclear – for example West German car manufacturers appear to be decreasing their foreign components, whilst some of their electrical component suppliers may be encouraging the use of forms of Kanban in their overseas plants. The location of Saturn in Tennessee is widely thought to have been chosen partly to allow GM to utilize the supply sources that Nissan has been developing for its US plant. Whatever judgement one makes of the compatability of Kanban with international sourcing – and the corporate strategists whom I interviewed saw no fundamental conflict – we should not underplay or take international developments for granted. Key actors in the development of new labour relations in the industry have themselves linked this to internationalization, as they are seen as ways of keeping jobs in-house and in the US (see, for example, Kertesz, 1986). However, in recognizing the importance of the internationalization process we must break its association with product standardization and the transfer of all low-skilled jobs to the newly industrialized countries, except in so far as component manufacture accounts for a disproportional amount of such jobs. Internationalization is always likely to be greater in components and design than in final assembly.

Finally, there is the role of uncertainty. Managements may genuinely be uncertain about future developments in technology and the link between these and the demands that will be made on workers. Luria (1986: 23–4) shows, for example, that the management of at least one GM plant, having collapsed the demarcations between the skilled trades, found itself reverting to some differentiation between electrical and mechanical skills largely for training reasons. The current concern managements are showing for flexibility does not reflect the advent of the new world portrayed by Piore and Sabel, but a desire for increased worker input to increase in particular the speed at which major new products can be introduced. A particular car plant may be relatively constrained in terms of the variety of products it can manufacture at any one point in time, but managements must plan for quick product changes and retooling times. Technological change is treated as a continuous process – the new technology both has evolved and is evolving – and managements' often rather loose use of the term flexibility relates to their desire to retain the freedom to adopt new technology, which itself might or might not involve shifts in the nature of the product. So, underlying the national programme of worker participation in the US auto industry which involves plants which have not yet radically changed technologically, or may even close in the next decade, is an anticipation that the successful adaptation of future technological possibilities will be achieved more easily with what managements see as an involved workforce. As the much publicized poor performance of GM's new plant in Hamtramck Detroit (Fisher, 1986: 56–7) reminds them, managements also anticipate being faced with technical problems which may require considerable debugging, when worker input may be vital.

As part of coping with uncertainty, managements will attempt to some extent to control their product markets. It is important then to stress that the product market, assumed in the flexible technology thesis to be increasingly fragmented, is itself not independent of the companies' marketing strategies. The trend would appear to be towards a 'defined variety' (Galjaard, 1982: 51) – defined by the companies – and not a substantial increase in variety rooted in developed flexible production processes. Furthermore, the environment to which certain workers are being asked to adapt in the name of market forces is increasingly populated by organizations (such as Saturn) and trading relationships (joint ventures) which their own senior managements, and elements of their union, have helped to create.

Conclusions

It seems possible to identify three eras of the participation debate in the US auto industry which we may for convenience term (1) the human relations (blue-collar alienation) stage; (2) the crisis (quality and productivity improvement) stage; (3) the technological (Saturn–After Japan) era. As we saw earlier, Katz (1985) suggests that managements became more committed to worker participation in the late 1970s as they began to define it as a way of reducing the influence of the contract and the use of the grievance procedure. Initially, then, worker participation was to modify the Fordist system. The schemes appear to have taken on an enduring quality in the 1980s with the intensification of Japanese competition, use of new production methods, and transition to higher levels of automation. In the third era plants are increasingly assumed to run on the basis of teams and hence in a permanent problem-solving and group mode. The worker participation schemes

can cater for a variety of specific local concerns and activit
general they aim to adapt workers to the new working ar
which call in most cases for more involvement, responsibility
possibly more abstract thought by workers. Furthermore, the restructuring of the industry, and the uncertainty surrounding this, entails the
reconstitution of its workforces.[13]

The formal participation schemes are, as Sabel (1982: 213) implies,
in a sense transitional since the new plants have no formal programmes
and are constructed on the basis of team working. There does not,
however, seem – at least at this stage – to be an underlying
transformation towards flexible specialization. What we have witnessed
in the US auto industry is a relative increase in flexibility and
particularly the elimination of some of the rigidities of Fordist methods
as conventionally conceived. A new era of mass production is, nevertheless,
developing with increasing computerization and integration between
functions. It is, however, premature to talk of a transformation away
from Fordism and certainly of a discernible trend towards flexible
specialization; treating the new era as falling in the middle ground
between Fordism and flexibility may then be misleading.

Even if one can delineate certain trends in the overall development
of the auto industry, it may be a mistake to draw firm conclusions
about overall tendencies in labour relations. Because of the wide
variation in the tenor of labour–management relations between plants,
especially within GM in the early 1980s – noted for example by Katz
(1985: 128) – several commentators have indeed questioned the extent
to which the companies are pursuing a coherent policy. This problem
is compounded by the fact that worker participation[14] is of course only
one part of managements' overall personnel strategy, which also
includes changing the systems of compensation, increasing the power
of the supervisor relative to the union committeeman and the reliance
on the grievance procedure, increasing direct communication between
the worker and the management, and maintaining managements'
prerogative over investment, location and sourcing decisions. If
managements' strategies were moving in a reasonably straightforward
direction, then the differences between plants, for example between
the new Fiero-type plants and the older plants, could be treated as a
matter of degree. If, however, the less co-operative – traditional, as
Katz (1985) calls it – approach to industrial relations dominates in
some plants or even at corporate level thinking, then there may well
be a qualitative difference between situations. We will then be tempted
to conclude that firms are following a variety of strategies or at least
a multi-stranded approach which may contain many contradictory
elements.

relations strategy, it would appear that, even in the third era, the
flexibility sought from workers bears little resemblance to that of the

multiskilled artisan which is implied by Piore and Sabel's and others' somewhat loose use of the old craft model with its overemphasis on autonomy and the transferability of skills. Furthermore, most managerial prerogatives remain firmly intact and the exact increase in the amount of discretion given to the shopfloor in the new regimes remains to be researched. The schemes are mainly confined to the operational level; they are oriented towards very immediate managerial goals; and they are under the control of management in so far as the implementation of workers' ideas depends on the 'normal channels' (GM plant manager). Above all else, while forsaking an element of their prerogative over some decisions, managements have maintained their flexibility in deciding on investments, technology and sourcing.

This chapter has emphasized the development of the schemes rather than evaluating either them or the reactions they have engendered. Virtually no independent evaluations are as yet available. Katz (1985) compared the experience of QWL in various GM plants in the late 1970s and concluded that the programmes (which were at the time less advanced than some of those I have discussed above) had a modest positive effect on industrial relations. The problems of quantitative evaluation[15] include access to data and the fact that every plant is starting from a different position. Moreover, worker participation is likely to be introduced sequentially, that is, initially in plants with a relatively good industrial relations record, then in plants with a bad one and where improvements are most feasible or most needed, and only after this is it encouraged elsewhere. Thus, even comparisons of rates of change in indices may not suffice to assess the impact of employee involvement schemes. There is also the vexed question of attributing the cause of changing performance to one particular programme, especially when some of the plants being researched are threatened with closure.

The objective of attitudinal structuring underlying the schemes adds to the evaluation problem. Seemingly minor symbolic changes – the single status canteens and car parking facilities for example – must be judged against the aim of increasing commitment which is essentially a matter of attitudes. It may also be very difficult, in the absence of participant observation, to ascertain what is happening to some of the more unobtrusive measures of the traditional adversarial relation, for example, relatively hidden practices such as doubling up, sabotage and cutting corners. It is also not clear whether or not the supervisors when faced with production difficulties, will revert to their old authoritarian ways of communicating with the worker. Such issues may perhaps be even more difficult to unravel in the climate of participation because to discuss them is to admit that they existed in the past. My own direct observations suggest that the majority of managements do genuinely support the programmes. They are committed to spending money on them, support the idea that workers' input is valuable to the production process, and are strongly committed to strengthening the co-operative elements in labour relations. Though difficult to measure, there are many visible signs of changes in the plants – for example the more relaxed and less authoritarian and assertive body language of managers towards workers.

Partly because of the difficulties of making evaluations and partly because of differing perspectives, there exist differing judgements of the effectiveness of worker participation (cf. Katz ch. 13). Consequently, it is especially important to stress the point, often made in the labour process debate, that the effects of managerial initiatives are not pre-established and that workers' responses will shape their outcomes. The issue for the union was initially formulated as largely one of monitoring the situation to ensure that worker participation groups did not encroach on those elements of the agreements which were jointly determined. But the growing strength of worker participation as a possible institution in opposition to the union, is coming to the forefront of debate amongst active trade unionists, allied as it is to their fears about Japanese-style 'enterprise unionism'. Thus particularly with the advent of NUMMI and Saturn, the issue for the union is again the precise role they are being accorded in managerial strategy. This is, of course, why, despite the joint nature of the programmes, worker participation has not as of 1985 been endorsed as official UAW policy.

Nothing in what I have said is intended to convey that a new consensus, or Japanese-style enterprise unionism, automatically results from the initiatives I have surveyed. Certainly, the conflicts surrounding the schemes will affect their future development. Moreover, the elements of the 'new' industrial relations which are concerned with increased managerial control may themselves become a source of conflict. The depoliticized and consensus model underlying modern human resource management and Saturn may not be achievable. But the aim, at least from management's point of view, is to change the rules of the game, or to substitute one game for another – the expectation being that the inevitable contingencies of production as well as the continual search for improvement will define shopfloor controversy. Whilst this may be interpreted by some managers and unionists as implying that in the plants of the future the union is expected to play a less active role in defining the terrain of contest on the shopfloor, worker participation should not be seen as inevitably weakening the union, or reducing the long-term rewards of workers.

There may, however, be sufficient overlap between the interests of the large-scale corporation and workers for such schemes to offer some increased scope for job satisfaction, respect and participation for workers. But any transformation of these initiatives into a broader movement towards industrial democracy would appear to depend on workers, presumably through their union, developing their own set of independent objectives. Part of the UAW's strategy surrounding new developments such as Saturn is that it may be easier to organize struggle on the shopfloor because workers will have increased control and involvement at that level. Nevertheless, as Katz (chapter 13) suggests, the trade union movement is divided over the new developments. Concern about the role of the steward and the sourcing issue are major examples of controversies surrounding the Saturn project. It is the emergence of new standards, for example in manning levels and in higher import content, and the way they can quickly be presented as *faits accomplis*, which most alarms those within the union who are most critical of the new schemes. Concern has also been expressed about the UAW's willingness to agree terms and conditions for a workforce ahead of its recruitment (Serrin, 1985). More generally, it is important to stress that the perspective of many unionists reflects the uncertainties either inherent in the situation or in the managerial strategies to which they are responding. For example the position of the majority of the union officials I spoke to at the Rouge local in Dearborne (Detroit) in effect fell between Katz's co-operatives and militants, as they could see both the benefits of worker participation and its costs, particularly its potential to undermine the union at plant level. They were genuinely uncertain as to whether they should kill off worker participation, push it to its limits, or simply be indifferent to its future development.

To conclude on the issues with which I began this chapter, there seems to be a need to recast the debate: control and flexibility should not be treated as two poles of a single continuum. On theoretical grounds there is a need to move away from the simplistic notions of scientific management and its association solely with Taylorist methods of factory management. The emerging production system seems very much a product of a scientific approach to management. The research in this chapter points to the way in which both the industrial relations and corporate strategies of the large auto companies are following complex paths which appear to weave between the two extremities of Fordism and flexible specialization which academic debate has constructed. It is as much the realities of the middle ground, as any theoretical inadequacies in labour process theory and Piore and Sabel's model, that demand new concepts. 'Between Fordism and Flexibility' is best studied without the clutter of iron laws of capitalism or romanticized visions of a return to craft work.

Notes

This chapter has benefited greatly from my involvement in the MIT 'Future of the Automobile' programme – I was a member of both the British and industrial relations team. It particularly benefited from my discussions with Knut Dohse, Harry Katz, Uli Jürgens, Tim Morris, Wolfgang Streeck and James Womack. I would also like to thank Fred Block, John Calvert, Ben Harrison, Dan Luria, Bryn Jones and Stephen Tolliday for comments on an earlier draft. The fieldwork was carried out between March and July 1985. It involved interviews with key representatives of the industrial relations and corporate strategy managements of the main car firms in the US and with members of the departments of the United Automobile Workers (UAW) dealing with these companies, including the vice president, as well as members of the research department. It also included nine visits to plants with varying degrees of worker participation, including two which have been given a great deal of publicity, GM's Tarrytown plant and the Livonia Engine plant. These visits included interviews with personnel managers, union representatives, worker participation functionaries, and in some cases shopfloor workers. In all cases I made a factory visit. I also had a three-hour presentation and discussion with five key members of the Saturn joint working party. I have also drawn on documents and newspaper articles, as well as the documentation gained through my involvement in the 'MIT Future of the Automobile', programme. Finally, I would like to thank the Nuffield Foundation and Fulbright Commission for their financial support for my fieldwork.

1. Kelly (1985) shows how new forms of job design can increase workers' control over, for example, their pace of work, whilst simultaneously increasing elements of management's control. This point has also been noted in the case of the now infamous Volvo experiments in the Swedish car industry; for, whilst the radial system of organization increased the autonomy given to the work groups, jobs remained timed, and other forms of direction simultaneously increased management's total production control.

2. See Katz (1985: 72–4) and Roberts *et al.* (1979: 59–61) for brief accounts of the early participation initiatives.

3. Quality circles also need not be viewed as inherently co-operative; several managers I interviewed in late 1985 in engineering in Brazil were reluctant to introduce them on the grounds it would foster increased discussion and solidarity between workers, and by implication might increase the salience of conflictual issues and unionism.

4. Pay for knowledge is a system of payments whereby people are encouraged to learn and train for a variety of tasks, and the learning is reflected in their pay packet.

5. In some situations there will be differing degrees of commitment to the programmes. There are quite intense differences of opinion in some plants about whether the worker participation should be supported, and in some cases these divisions may reflect deeper, more long-standing political differences. Also of considerable importance are the differences in the potential for participation which groups may have within the same plant. So, even in one of the plants built in the 1980s with a very strong participation scheme, I observed differences in the implications of the programme for different groups. For example, on the final inspection section, it fostered genuine group working; for many of the male production workers, who were working individually, the scheme had basically

facilitated regular weekly small group (quality circle-type) meetings. For the women who were working on short assembly lines at various points in the production process, it largely meant job rotation, that is increased mobility between very limited tasks, albeit on a basis worked out by the women themselves.

6. Lee Iacocca, the chairman of Chrysler, for example said, 'We are going to go wherever we have to go to get quality parts at the best price' (Mateja, 1985: 36). Roger Smith, the chairman of GM, in a letter to all GM employees dated 25 February, 1982, said, 'every day we must decide whether to make a component ourselves or to buy it from an outside supplier. The fundamental laws of business demand that, where quality and delivery are equal, we must go to the source where cost is lowest.'

7. As the share of imports in the total US market has been increasing, so too has the Japanese share of the imports. In 1960, their cars represented 0.4 per cent of the total imports, a figure which rose to 79.6 per cent in 1980, and continued to increase to 82 per cent in 1982, despite the introduction of import quotas in 1981. The sheer scale of this quota, fixed at 22 per cent of the total US market, reflects the massive growth of imports from 1978 onwards when they represented about 4 per cent of the total car sales. Estimates of productivity growth point to the considerable rise in Japanese labour productivity, at a time when the US productivity was falling. For example, Altshuler *et al.* (1984: 203) produce an index of units produced per hour worked in the years 1980–1 compared with 1970–7, which for Japan is 166, when the comparable figure for the US is 91.

8. Quotation from a speech by the chairman of NUMMI, Mr Toyoda, forwarded to the author by the company.

9. Robots are thus far mainly confined to routine semi-skilled tasks, such as welding and paint spraying. But the use of robots is increasing rapidly; Chrysler for example had 16 robots in 1973, whereas their new Sterling Heights plant (near Detroit) alone has 57 welding robots, 35 material handling robots, and 162 lasers and cameras for inspection.

10. Jones (1986) points to the rather limited nature of the flexible systems even in Japan: 'Japanese FMSs are not particularly notable for being aimed at securing qualitative gains or more product variability'.

11. Capital is lumpy and the firms have heavy investment in existing facility. So, as Unterweger (1983), of the research department of the UAW puts it:

> The great productivity gains achievable by integration will be realized gradually in existing facilities. New facilities will most likely show a much higher degree of integration and productivity. . . . It is probably easier to design highly integrated plants from scratch than to convert existing facilities. In short, the full effects of integrated factory automation will not be felt until the existing capital stock has been turned over.

12. In reaction to Katz's and Sabel's (1985: 297) judgement that, 'the fluctuations of general world trade . . . undercut the potential benefits of the world car strategy', Ford World Corporate Management wrote (personal communication to the author) thus:

Ford Motor Company did reap benefits from the Escort in the areas of reducing time to develop the product; more efficiently utilizing the product development process and improving the quality, both in Europe and the United States. The Escort was a world car and the differences between the European and the US models were minimal. [But] it is difficult to generalize that the corporate strategy of Ford is the world car.

13. As Lazonick (1983) and Johansson (1986) suggest, Taylorism and Fordism were concerned with constituting new 'semi-skilled' workforces and not largely, as others have stressed, with the decomposition of the skilled workforce.

14. Corporate managements do not therefore judge a plant's labour relations development simply by its worker participation. There have in fact been instances of investment in plants without developed worker participation schemes. The Kelsey-Hayes, Romulus, plant in Michigan, for example, was given a new product line and new aluminium spinning technology for aluminium wheels in 1985.

15. No independent study, so far as I know, has been carried out to assess the reaction of workers to the programmes. Reports in union publications such as those of the Society of Engineering Office Workers which represents workers in Ford's body and design departments in Detroit tend to highlight the negative aspects. For example there are criticisms that management does not fully support the programme, and about the lack of perseverance on everybody's part and the insufficient authority given to the worker participation groups.

References

Abernathy, W.J. 1978. *The Productivity Dilemma.* Baltimore: Johns Hopkins.

Adler, P. 1985. 'Managing Flexibility: A Selective Review of the Challenges of Managing the New Technologies – Potential for Flexibility.' A Report to the Organization for Economic Co-operation and Development, mimeo. Stanford: Stanford University, Dept. of Industrial Engineering and Engineering Management.

Altshuler, A., M. Anderson, D. Jones, D. Roos, and J. Womack. 1984. *The Future of the Automobile.* London: Allen & Unwin.

Dankbaar, B. 1984. 'Maturity and Relocation in the Car Industry'. *Development and Change.* Vol. 15, no. 2, 223–50.

Dohse, K. 1986. 'Konzern, Kontrolle, Arbeitsprozess'. *Prokla.* Vol. 62, March, 74–104.

Dohse, K., U. Jürgens, T. Malsch. 1984. 'From "Fordism" to "Toyotism"? The Social Organization of the Labour Process in the Japanese Automobile Industry.' Wissenschaftszentrum Berlin: Discussion Paper IIVG/pre84–218.

Dombois, R. 1986. 'The New International Division of Labour, Labour Markets and Automobile Production: the Case of Mexico'. *The Automobile Industry and its Workers,* Ed. S. Tolliday, and J. Zeitlin, Oxford: Polity Press, 224–57.

Fisher, A. 1986. 'GM is Tougher than you think'. *Fortune.* 10 November, 56–64.

Friedman, D. 1983. 'Beyond the Age of Fordism: The Strategic Basis of the Japanese Success in Automobiles'. *American Industry in International Competition*. Ed. J. Zysman, and L. Tyson. Ithaca: Cornell University Press, 350–90.

Fröbel, F., J. Heinrichs, and O. Kreye. 1980. *The New Industrial Division of Labour*. Cambridge: Cambridge University Press.

Galjaard, J.H. 1982. *A Technology Based Nation*. Delft: Interuniversity Institute of Management.

Guest, R.H. 1979. 'Quality of work life – learning from Tarrytown'. *Harvard Business Review*. Vol. 55, July–August, 76–86.

Guest, R.H. 1983. 'Organizational Democracy and the Quality of Work: the Man on the Assembly Line'. *International Yearbook of Organizational Democracy*. Ed C. Crouch, and F.A. Heller. New York: Wiley, 139–54.

Harvard Business Review. 1981. 'The Automobile Crisis and Public Policy'. Interview with P. Caldwell, Vol. 59, January–February, 73–82.

Hayashi, M. 1982. *The Japanese Style of Small-Group QC Circle Activity*. Tokyo: Chuo University.

Hill, R.C. 1983. 'The Auto Industry in Global Transition'. Paper given at the Annual Meeting of the American Sociological Association, 3 September, Michigan, Detroit.

Holusha, J. 1986. 'A New Way to Build Cars.' *New York Times*. 13 March, D2.

Johansson, A. 1986. 'The Labour Movement and the Emergence of Taylorism'. *Economic and Industrial Democracy*. Vol. 7, November, 449–86.

Jones, B. 1986. 'Cultures, Strategies and Technical Essentials'. Paper given at the International Workshop on New Technology and New Forms of Work Organisation. Berlin (GDR): Vienna Centre and Nationalkomitee für Soziologische Forschung.

Katz, H. 1985. *Shifting Gears*. Cambridge, Mass: MIT Press.

Katz, H., and Sabel, C. 1985. 'Industrial Relations and Industrial Adjustment in the Car Industry'. *Industrial Relations Journal*. Vol. 24, Fall, 295–315.

Kelly, J. 1985. 'Management's Redesign of Work: Labour Markets and Product Markets'. *Job Redesign: Critical Perspectives on the Labour Process*. Ed. D. Knights, H. Willmott, and D. Collinson. Aldershot: Gower, 30–51.

Kertesz, L. 1986. 'Bieber, Warren View Challenge of Industry's Globalization'. *Automotive News*. 6 October, 13.

Kochan, T., H. Katz, and H. McKersie, 1986. *The Transformation of American Industrial Relations*. New York: Basic Books.

Lazonick, W.H. 1983. 'Technological Change and the Control of Work: The Development of Capital–Labour Relations in US Manufacturing Industry'. *Managerial Strategies and Industrial Relations*. Ed. H.F. Gospel, and C.R. Littler. London: Heinemann, 111–36.

Luria, D. 1986. 'New Labor–Management Models from Detroit'. *Harvard Business Review*. Vol. 64, September–October, 22–8, 32.

Malsch, T., K. Dohse, and U. Jürgens. 1984. 'Industrial Robots in the Automobile Industry, A Leap Towards "Automated Fordism?" Wissenschaftszentrum Berlin: Discussion Paper IIVG/dp, 84–222.

Manwaring, T., and S. Wood, 1985. 'The Ghost in the Labour Process'. *Job Redesign: Critical Perspectives on the Labour Process*. Ed. D. Knights, H.

Willmott, and D. Collinson. Aldershot: Gower, 171–96.

Mateja, J. 1985. 'We have to get a global view'. *Jobber and Warehouse Executive.* September, 35–7.

Maurice, M. 1986. 'Flexible Technologies and Variability of the Forms of the Division of Labour in France and Japan. Paper given at the International Workshop on New Technology and New Forms of Work Organisation. Berlin (GDR): Vienna Centre and Nationalkomitee für Soziologische Forschung.

Nomura, M. 1985. "Model Japan"? Characteristics of Industrial Relations in the Japanese Automobile Industry'. Wissenschaftszentrum Berlin: Discussion Paper IIVG/d85–207.

Piore, M., and Sabel, C. 1984. *The Second Industrial Divide.* New York: Basic Books.

Reich, R.B. 1983. *The Next American Frontier.* New York: Times Books.

Roberts, B., H. Okamoto, and G. Lodge, 1979. *Collective Bargaining and Employee Participation in Western Europe, North America and Japan.* New York: The Trilateral Commission.

Sabel, C. 1982. *Work and Politics.* Cambridge: Cambridge University Press.

Schoenberger, E. 1985. 'Multinational Corporations and the New International Division of Labor: Incorporating Competitive Strategies into a Theory of International Location.' Baltimore: Johns Hopkins University Press, Mimeo.

Serrin, W. 1985. 'Saturn Labor Pact Assailed by a UAW Founder'. *New York Times.* 26 October, A13.

Shaiken, H. 1985. *Work Transformed.* New York: Holt, Rinehart and Winston.

Stock, R. Jr. 1980. *Management Or Control? The Organizational Challenge.* Bloomington and London: Indiana University Press.

Streeck, W. and A. Hoff, 1983. 'Manpower Management and Industrial Relations in the Restructuring of the World Automobile Industry'. Wissenschaftszentrum Berlin: Discussion Paper IIM/LMP, 83–35.

Tolliday, S. and J. Zeitlin, 1987. 'Introduction: Between Fordism and Flexibility'. *Between Fordism and Flexibility.* Ed. S. Tolliday and J. Zeitlin. 1–25.

Unterweger. P. 1983. 'Work, Automation, and the Economy'. Paper given at the American Association for the Advancement of Science Annual Meeting, 27 May, 1983. Detroit, Michigan.

Vernon, R. 1974. 'The Location of Economic Activity'. *Economic Analysis and the Multinational Enterprise.* Ed. J.H. Dunning. New York: Praeger.

Walton, R.E. 1985. 'From Control to Commitment in the Workplace'. *Harvard Business Review.* Vol. 63, March–April, 77–84.

7

New Technology in Scotbank: Gender, Class and Work

John MacInnes

Introduction

This case study of a bank offers a corrective to the male manual manufacturing bias of some studies of industrial relations and technical change because it examines non-manual work in the services sector done mostly by women. 'Scotbank' was examined between 1981 and 1984 as part of a longitudinal analysis of industrial democracy reported in Cressey *et al.* (1985). MacInnes (1986) is a longer account.

Background: The Internal Labour Market in Banking

Authority in banks is centralized to keep strict control over large sums of money, with 14 volumes of branch practice from head office determining how work is to be done. This centralized structure provides a range of staff positions differentiated by seniority (from junior clerical staff to the control of head office departments) and by function. These posts are filled almost entirely by promotion from within, as banks grow their own 'experts', producing all round 'general' bankers from school leavers. Seniority is not simply a hierarchy of offices. Many staff have 'management' functions, while senior branch managers themselves will be subject to higher authority. Managers are both agents of head office and its authority in the branch (including industrial relations issues) and representatives of branch staff to the bank, recommending promotions or pressing for higher staffing levels, for example. Both people and posts are graded, with recruits entering the bank on grade 1. After 1–2 years training both in branches and at the bank's college, staff gain a certificate of competence and progress to

TABLE 7.1
Scotbank 1982

Grade	% of all Clerical Staff in that Grade	% of Staff in the Grade who are Male	% of Staff in the Grade who are Part-time Females
1	16	18	64
2	43	23	—
3	13	38	—
4	8	56	—
5–9 (accountants)	12	88	—
10–14 (managers)	8	99.5	—
All staff	100	41	9

grade 2: the general clerical grade. Over half the bank's staff are on these two grades. Those who intend to make a career in banking study for the Institute of Bankers (IoB) exams. On passing these exams employees automatically qualify for promotion to grade 3 (bank officer) and may then be promoted to grade 4 (bank supervisor). Promotion to these positions also involves further training. Employees with IoB qualifications may be promoted to 'appointed' positions (grade 3+) which are more managerial in character (overtime, for example, ceases to be payable). Accountants usually administer branches, while managers concentrate on business development. Table 7.1 shows the distribution of staff in the grades in 1982.

The Organization and Methods Department monitors branch work-loads through the Clerical Work Improvement Programme (CWIP) and grades posts accordingly. Promotion from one grade to another involves therefore a combination of the person being qualified and a post being available. It therefore requires considerable geographical mobility:

> When you're transferred or moved in the bank you're told you're going and if you refuse then your chances of promotion are that bit less . . . you're probably snookered for a good bit. But then again you know that when you join the bank. You've got to be on the move and you've just got to put up with it (Staff interviews).

Staff are appraised annually on the basis of their individual performance. Thus their salary is a combination of the grade of their post, length of service in that post, and appraisal rating.

Because of the complexity of matching posts (determined by business strategies, marketing, organizational structures and technologies) to people (determined by recruitment, training policies and turnover) in the internal labour market, the role of the staff department is central. Moreover, staff costs form about three-quarters of a bank's expenditure. Just as the staff department is central to the bank as a whole, the bank's business strategy more directly affects the jobs and careers of its employees. New functions or services may affect not only the job content, but the pattern of employees' careers; perhaps fewer branch managerships, more head office positions, or posts overseas or in different areas of the UK, and so on.

Because of the internal labour market all new staff are potentially career bankers but not all recruits expect or want to pursue such a career or study for exams. There is, therefore, a distinction to be drawn between bank staff there for a 'career' and for a 'job'. Careers produced by the internal labour market are central to the banks' approach to industrial relations. It promises not only job security but eventual high earnings and responsibility to compensate for lower pay and greater mobility earlier on. Low-interest mortgages facilitate the geographical mobility of staff and also make leaving the bank potentially expensive. It is also very difficult to leave one bank and join another. Most resignations are by clerical staff who have not followed a career in the bank; about 10 per cent of grade 1 and 2 staff left in 1980, compared with under 1 per cent of appointed staff.

Only in the labour shortages after the Second World War did banks begin to recruit women and then only as clerkesses. Blackburn (1967: 71–3) commented 'careers in banking are for men; the routine work such as machine operating is for women. . . . Quite openly, they regard men and women as two different classes of employees.' Since then, banks have introduced equal opportunities policies and promotion has been open to all, but in practice female staff are concentrated in the junior grades, are less likely to take professional qualifications and are more likely to leave. While the banks argue that there are now equal opportunities for men and women, the union's equal opportunities committee argues that more positive action is required because women are less well placed than men to take advantage of these opportunities. They may decide to follow a career relatively late, or may find the requirements of geographical mobility harder to fulfil for domestic reasons. In our attitude survey of a sample of Scotbank employees, 90 per cent of male unappointed staff thought they had a very or fairly good chance of reaching an appointed post, but only 26 per cent of females. The figures in table 7.1 suggest the women were being optimistic. The bank argues that most women aren't interested in a career because of the commitment to training and geographical mobility required, and that most will leave to start a family. The union argues that these are self-fulfilling prophecies: it is assumed that boys will study for banking exams, girls have to overcome resistance (informal and unconscious as well as explicit) to do so. Moreover it is more difficult to take exams later, which affects older women who decide to pursue a career.

Paternalism

The importance of security in handling money encourages an air of caution, sobriety and conservatism. Banks cultivate this image of trustworthiness to reassure customers. This influences the recruitment and training of staff who 'are expected to be courteous and attentive in their relations with customers, the public and other members of the staff, and their general bearing and appearance should reflect credit, not only to themselves, but on the Bank' (*Scotbank Staff Manual*). Heritage (1983: p. 135) notes that in small branches 'flexibility and mutual co-operation among staff are essential for the smooth running of the branch. The atmosphere of secrecy and, in modern times, danger which attends large financial transactions promotes a sense of "teamwork" and solidarity.' Along with the internal labour market, financial benefits like profit-sharing and special mortgages, the social and educational background of recruits, this ethos is part of a corporatist relationship between the bank and its employees, especially career staff, which is seen as one which extends beyond instrumental calculation of mutual economic advantage. The bank offers a life-long career and expects in return active loyalty. This corporatism was real but limited. Collective industrial conflict was rarely threatened and almost never used, but there were keenly felt conflicts of interest, grievances and disagreements. Moreover many junior clerical staff, especially women, had little expectation of any career in the bank.

Economic and Technical Changes

Banks have been using new electronic technology for over 20 years, but by 1980 technological innovation was giving rise to new services (such as 24-hour 'through the wall' cash dispensers and home banking) as well as automating old ones. It was also permitting 'satellite' banking and 'twinning' of separate branches under one manager. 'CHAPS', a system for the automatic same day clearance of cheques, and 'EFTPOS', the provision of direct bank to retailer fund transfer were being developed. Some estimated that such innovations would reduce employment by 10 per cent by 1990 in contrast to previous substantial expansion. Meanwhile the Conservative government was deregulating the finance sector, increasing competition with building societies, insurance companies and City institutions. International and domestic bad debt problems increased dramatically because of the economic recession.

These changes in markets and technology have affected recruitment, training and careers. Although banks still 'grow their own' staff to be all-round bankers, the division of labour has changed, diluting the tradition of 'generalism'. Unqualified young people are employed for cheque-clearing work: an assembly line-like process. Specialist graduates are now recruited too, particularly in electronic data processing (EDP). The need to attract computer specialists has disrupted the banks' internal labour market. The division between jobs and careers in banking has become harder. The single, long avenue of promotion of the past in which almost every employee had a potential interest has been replaced by a range of more specialist career paths involving management training programmes and more experience in head office for young 'high flyers'. The increasing volume of business and extended range of services which banks offer, and range of customers they deal with, has increased the standardization of branch working practice and also the workload on appointed staff. They have less discretion over how to operate a much wider range of services which they must master and market. This changing organization of work in banking might be described in terms of a shift from 'paternalism' to 'technocracy': aristocratic figureheads in the boardrooms have been supplanted by EDP specialists. The 'all round' banker has been fragmented into several highly technical specialisms. But this process has limits. Banks remain conservative, it is in-house training that still predominates, and all round experience is still valued since it softens departmental frictions by giving career staff wider experience.

New 'daily staffing' systems have been introduced too, whereby banks employ part-time (overwhelmingly female) staff in clerical jobs to staff branches at peak business periods in the week. Greater competition between banks and building societies has also threatened a return to six-day opening. This issue raised strong feelings particularly among older staff. The campaign for a five-day week in the 1960s was very significant in drawing members into the Banking Insurance and Finance Union (BIFU) and giving it national bargaining power. Related issues were opening hours, hours of work and the length of the working week – BIFU had negotiating rights on the last of these but not the first two. There was considerable debate within the union about the best way to trade off a potential shorter working week against changes in hours of work, and over the extent to which competition made changes in bank opening hours inevitable.

Industrial Relations

At the time of the research Scotbank negotiated terms and conditions nationally for clerical staff and at company level for appointed staff with BIFU. Five sixths of the men and two thirds of the women in Scotbank were members. Negotiation, consultation and communications in the bank were organized centrally, reflecting the link between staff matters and decision-making outlined above, and uniformity of conditions across branches required by staff movement. There is no local bargaining and little that can arise at branch level beyond individual grievances over interpretation of the staff manual. There were no lay union representatives in the bank's branches, and BIFU branches were geographically organized. Scotbank employees in these branches elected delegates to the Scotbank Institutional Committee which included the full-time official responsible for Scotbank, and the seconded representative, a full-time lay official. Full-time union officials played an important role in consultation, bargaining and other industrial relations activity. Lay officials would neither formulate policy nor negotiate without consulting the full-time officials or having them present.

In addition to BIFU there were three other staff organizations in Scotbank. The Accountants Association organized social events at which there was a formal business session and a guest speaker, and met four to six times a year to discuss issues of interest to accountants such as bank operating practices or promotion opportunities. The Managers Association was a parallel body for managers. Neither body was permitted to raise issues for which BIFU was recognized. The Head Office Officials Consultative Group was a body similar to the two other associations which represented staff of managerial grade in head office on the bank's Joint Consultative Committee (JCC).

Relations between BIFU and Scotbank reflected the corporatist ethos outlined above. As an Institutional Committee member described it:

> We're not anti the management, we're probably in a sense pro the management because we want to see ourselves advancing, we want to see the bank advancing. To my mind unionism in the bank is different from unionism in the industrial sense. We want to see things moving forward and we are prepared to come and go a lot which you wouldn't get in outside industry.

But the contrast between industrial relations in the bank and elsewhere can be overdrawn. The attitude survey showed that the vast majority of staff and union representatives were 'pluralist' in their outlook (MacInnes, 1986). Bank staff have taken limited strike action in the past over low pay, opening hours and hours of work. Scotbank enjoyed a reputation for enlightened industrial relations and in 1980, in part because of the degree of changes facing it, the bank established a new joint Consultation Committee which was to meet regularly and discuss papers submitted from the bank and the various staff bodies. The material conditions for developing industrial democracy in Scotbank were very favourable. The workforce was highly educated and articulate, less heterogenous and sectionalist than elsewhere, organized in a single union with a co-operative attitude towards the employer. The internal labour market and generalist training gave most employees an all round knowledge of the bank. Centralized bureaucratic decision-making particularly facilitated the implementation of decisions reached jointly.

New Technology in Scotbank

Scotbank was one of the first banks to introduce automatic cash dispensers, first using electro-mechanical machines and introducing new electronic cash dispensers in 1977. By mid-1980 some 100 machines were installed and at that time it was expected that by 1985 200 machines would handle half the bank's cash withdrawal. In practice the cash dispensers proved to be far more popular than anticipated so that the pace of introduction substantially exceeded original expectations. At the same time automated teller machines (ATM) were introduced between 1979 and 1981, removing about half the work effort in a deposit account transaction and providing new information to the teller. Back office automation proceeded more slowly. Branch working practices were extensively revised in order to apply the work effort released by cash dispensers and ATMs to other duties. Establishment levels for branches were changed. Peaks in the workload through the week produced by branch counter services were to be met by using part-time staff on a daily basis.

The introduction of this new technology neither changed working practices nor was the recruitment of part-timers negotiated with BIFU. Nor was there advance consultation. There was some limited discussion over the details of their operation and over payments for some new extra duties (e.g. call-outs to restock cash dispensers). This was despite a large education and publicity campaign by BIFU nationally to attempt to secure new technology agreements with the clearing banks. It produced a 'model agreement' which laid down negotiations over the planning, introduction and operation of new technology, revision of job evaluation structures and movement towards a 28-hour 4-day week, longer holidays and earlier retirement. Its attempts to secure such agreements came to nothing. It found little response even amongst the union's lay officials. There was no question of obstructing new technology pending negotiation of the terms of its implementation. Scotbank dismissed BIFU's case as alarmist, and denied any foreseeable need for compulsory redundancies. It emphasized its record of 20 years successful introduction of microelectronic new technology and the common interest of staff in maintaining the bank's competitiveness. It saw no widespread concern from staff about the implications of new technology which it needed to address, and was under no obligation through either domestic or national procedural agreements to negotiate or consult on the issue.

The Popularity of New Technology

Several factors seem relevant to explain this situation. The corporatist industrial relations tradition of banks outlined earlier militated against the development of opposition to the banks' plans. But the Scotbank workforce did not accept new technology because they thought that management ought to be free to make such decisions itself. In the employee survey 80 per cent agreed that 'it's management's job to manage' but only 20 per cent thought that 'management run the bank. I'm just here to work'. 72 per cent thought 'staff should be more involved in decision-making'. 84 per cent thought 'I would like to know more about changes before they happen'. When asked about industrial democracy, only one respondent thought it 'undesirable', half defined it as 'more consultation' and a third as 'better communications'. JCC members complained that consultation on new technology came too late.

More important was the context within which new technology was introduced and the immediate effects of its introduction for the staff working with it. The past record of the use of computers was relevant. It was not obvious to staff that the latest introduction of microelectronic technology should be qualitatively different from the last 20 years. Lay union representatives accepted the bank's arguments that new technology was important for competitiveness (though they were sceptical of just how strong the pressures on the bank were at a time when profits were at record levels). Introduction came at a time when business levels were rising at 6–7 per cent per annum. Pressure of work in branches could be quite intense and was increasing. The bank had a clear policy of restricting the growth of staff numbers as much as possible to restrain the growth of labour costs and avoid the prospect of redundancies if their demand for labour should slacken, since this would undermine one of the foundations of the industrial relations system in the bank: job security. New technology relieved this pressure of work considerably, especially cash dispensers which the public preferred to use. Moreover, rather than deskilling work, new technology automated more routine aspects like counting or checking. Finally, bank staff thought cash dispensers would help defend existing opening hours, because they gave an automatic 24-hour service. The employee survey found only 23 per cent who thought that 'new technology has not made much difference to my job', mostly appointed and head office staff who did not work so directly with it. Only 2 per cent claimed they had not seen it produce changes in other jobs. But the evaluation of new technology was positive. 73 per cent agreed that 'it's made my job easier' and only 6 per cent agreed that 'it's made my job more boring'. Most staff still thought their jobs involved 'a lot of responsibility', and only 12 per cent found them 'boring'. Ninety-eight per cent thought they had good job security, 86 per cent a friendly work environment, 77 per cent 'interest and variety'. Representatives on the JCC also took an overwhelmingly positive view of new technology; only 2 criticized its technical aspects at all. Thus an early discussion of cash dispensers on the JCC had union representatives pressing for their more rapid introduction. The union noted the 'rapid reduction in (staff) numbers' but did not press the point. The staffing situation was obscured by other changes. The most contentious point was a technical one about how staff were called out to refill machines.

The introduction of new technology coincided with substantially increased employment of grade 1 part-time staff which was cheaper for the bank. Scotbank did not discuss this with BIFU except to offer negotiating rights for these staff. These part-time workers were women – former bank employees who had left bank jobs because of domestic commitments, usually to raise a family. For security and training reasons this was the main source of 'peripheral' labour available. The existence of this pool of labour was in turn dependent on the gender basis of the career–job distinction outlined earlier and women's domestic subordination to their male partners. Only their inability to pursue a *career* in the first stage of their employment left these women available for part-time *jobs* in the second stage.

However, it would be wrong either to overestimate the bank's ability to expand the part-time labour force, or to make a direct association between part-time employment and the introduction of new technology: the events coincided but were not causally related. Part-time labour was cheaper with more flexible hours, but managers found such staff were less adaptable than full-timers, making work planning more difficult and less flexible. New technology aggravated this because it concentrated staff time on a growing range of customer services, sales and related back office functions for which part-timers were less suitable.

New technology reinforced the trend to specialization of function and centralization of control described earlier. Early service in head office, particularly on the management development programme became more necessary for senior promotion: a career in the branches was less likely to go so far up the promotional ladder. For established appointed staff, this appeared as a promotion blockage at the expense of younger 'high flyers'. Secondly the job of branch managers was seen to carry less prestige but also to be more onerous. Decisions about lending were more subject to formulae laid down by head office rather than the branch manager's discretion. However, there was instead a growing range of other services offered by the bank which the branch manager was expected to master and market. Moreover the additional range of information which new technology made available to him or her was supposed to be mastered and used in selling services and making decisions. It is not surprising then that for these reasons senior staff were less likely than their unappointed colleagues to think that new technology had made their job easier.

New Technology and Industrial Democracy

Set against the propitious environment for industrial democracy the JCC's achievements were marginal. Its main contribution to new technology was the technical point about monitoring cash dispenser refilling. Both sides expressed the hope that more important issues should be discussed but understood discussion in entirely different ways. The bank wished to inform employees, through the JCC, of why it was doing or had done various things and to speak in very general terms only of its future plans. The staff representatives were dismayed at what they saw as 'post mortems' and wanted more concrete information and discussion about the future, and complained that the JCC had achieved nothing concrete. The JCC's impact on staff was minimal. A large minority were unaware of it and few claimed to have seen its reports or talked about it with other staff. There was no evidence of any great staff desire to spend some of their free time discussing and proposing alternative futures for Scotbank. Either they felt that the current Scotbank management was proficient enough, or that their efforts to influence it (which in terms of time and resources would need to be great) would be unlikely to succeed.

Conclusions

I have tried to indicate why, in the case of Scotbank in the period concerned, the considerable agitational effort that BIFU put into its campaign over new technology and its attempt to secure negotiating rights over its introduction had little success. The bank's past record, the way it introduced new technology, its effects on the workload at branches, the lack of concern about employment effects in the context of an expanding market and hopes about opening hours all tended to make staff feel fairly positive about its introduction.

This does not mean that it was an issue without social concern. The bank's employees had secure jobs but job opportunities for school leavers and unemployed people declined in relative terms. The bank's anxiety to prevent rising staff numbers was perfectly rational for itself, but less clearly rational in the context of high Scottish unemployment. Second, new technology could have been used to facilitate other changes: for example shorter working hours or part-time work as part of a banking career. It could thus have served to erode rather than reinforce the sexual division of labour in banking. Why such discussion of these alternatives never surfaced is considered below. Third, except for some union activities, there appeared to be little discussion and concern over the health and safety implications of VDU technology. This is despite the claims of links between low-level VDU radiation and miscarriage in pregnancy, and the substantial numbers of women using such technology in banks.

Scotbank's industrial relations contrasted with those major features of British manufacturing held responsible for poor economic performance and inflexibility: craft traditions, sectionalism, multi-unionism, 'them and us' attitudes, insistence on managerial prerogative and high conflict. Since private services now employ many more than manufacturing, this should warn us against describing a British industrial relations system by features found in engineering factories. But Scotbank did not simply have better high trust relations. Unionists complained strongly about Scotbank's failure to consult meaningfully about new technology and keenly felt that the quiescence of bank staff cost them influence. Peaceful industrial relations at Scotbank was part of the corporatist ethos founded upon both job security and commercial security. It could not be transplanted to a world of fast changing product and labour markets found in manufacturing.

Technological change at Scotbank was not a shift from Taylorism to flexible specialization. These terms are a poor antimony because of the empirical ambiguity about what Taylorism is and because flexible specialization is itself a contradiction in terms. Most producers can only develop flexibility at the cost of specialization; it is the trade-off between the two which is vital. Scotbank was, and continued to be, Taylorist in that all branch staff closely followed procedures bureaucratically set out by head office. But learning these procedures and how to apply them in complex situations involved great amounts of skill and particular local knowledge which part-timers, for example, did not always have. New technology automated some of the most boring aspects of skill, it also increased central control and reduced local managerial discretion. But new services required new competences from staff. There was no deskilling in the manner described by Braverman (except in so far as branch managers' individual judgements about credit worthiness had less scope) nor was there a move to greater staff participation in organizing new forms of work which a flexible specialization analysis would imply. Finally the history of new technology at Scotbank shows it was a continuous process, not one of discrete dramatic changes. We should perhaps be cautious of waves or cycles of innovation which are too neat.

The Scotbank study supports other work on the complexity of skill which has emphasized the importance of gender (such as Phillips and Taylor, 1980 and Cockburn 1983). There was no clear class division in the bank between a deskilled proletariat and managers who designed and controlled: rather there was a continuum of posts with different responsibilities. But more important than a person's immediate grade were their future prospects: whether they had a job or a career. Since less than 2 per cent of staff were women in appointed posts this division virtually meant whether they were a woman or a man. The life chances and material interest of a male grade 2 clerk shared more with the grade 14 manager he might aspire to become, than the grade 2 clerkess at his side. In the arena of paid employment neither bank nor union recognized this division. Scotbank's interest was to stress that everyone was a potential career employee, while BIFU needed to emphasize the common rather than conflicting interests of its members. The division had roots in domestic unpaid work. Women's commitment to domestic labour, child rearing and geographical mobility required by their spouse's employment tied them to a job just as it freed their partners for a career.

For women there remained a vicious circle of subordination in paid and unpaid work. Their heavier share of domestic work prevented them competing equally even in a formally equal labour market. Their poorer returns and prospects in paid work then rationalized their greater share of unpaid work. But it was a division crucial to the social impact of new technology. In theory labour-saving new technology has been seen as a way to reduce hours of work. Unless this is done, protracted high levels of unemployment seem inevitable. Simultaneously long hours of paid work have been seen as a major obstacle to greater equality in participation in unpaid domestic work by men and paid work by women. But instead of breaking down the division in this way, new technology was part of a series of changes which reinforced it: through hardening the job/career split through management development programmes and the expansion of part-time labour. The alternative potential of new technology hardly got explored, a vigorous union campaign wielded little influence on the bank or its employees, despite the propitious environment for industrial democracy in Scotbank.

This was in large part because the links between paid and unpaid work had no formal expression. Although its campaign failed to win its stated objectives, BIFU's members' interests as immediate and future wage earners were well protected: most didn't even feel them threatened. As men who might resent the sacrifices a career would imply or as women who might henceforth leave their job to rear children then return only to part-time work, they had no representative. Conversely the bank was under competitive pressure to use new technology in the most immediately profitable way. It could not decide unilaterally that more widespread sharing of hours of paid work was a general social good.

We might conclude therefore that understanding economic innovation needs more than orthodox ideas of skill and of conflicts between labour and capital. These have to be supplemented if not supplanted, by concepts of gender and the relation between paid and unpaid work.

Note

I am very grateful for the generous co-operation I received from Scotbank and its staff, and the Banking Insurance and Finance Union (BIFU) Scottish Office.

References

Blackburn, R.H. 1967. *Union Character and Social Class*. London: Batsford.
Cockburn, C. 1983. *Brothers: Male Dominance and Technological Change*. London: Pluto.
Cressey, P., J.E.T. Eldridge, and J. MacInnes. 1985. *Just Managing*. Milton Keynes: Open University Press.
Heritage, J. 1983. 'Feminisation and Unionisation: A Case Study from Banking'. *Gender, Class and Work*. Ed. E. Gamarinkow *et al*. London: Heinemann, 131–140.
MacInnes, J. 1986. 'Participation in Scotbank'. Research Report, CRIDP, University of Glasgow.
Phillips, A. and B. Taylor. 1980. 'Sex and Skill: Notes Towards a Feminist Economics'. *Feminist Review*. Vol. 6.

Part III

Skills, Deskilling and Labour Market Power

8

Labour and Monopoly Capital

Peter Armstrong

Introduction: The Braverman Thesis on New Technology

The most systematic and general account of the fate of productive labour under capitalism remains that of Braverman (1974). Briefly, his argument is that, given the dictates of capital accumulation, capitalists and their managements are constantly driven to renovate the productive process. To date the principal methods for achieving this have been the battery of techniques associated with scientific management and the 'scientific/technical revolution'. Essentially both of these have enabled the conception of productive labour to be separated from its execution and appropriated by capitalist managements as an instrument of control. In this arrangement, the conceptual phase of productive labour now serves the additional function of a control mechanism whereby additional surplus value can be extracted from the workforce. Because of the greater potential of this arrangement for capital accumulation as compared to the unity of the labour process when left to 'craft controls', the development of capitalist economies thus tends to add 'real' control of the labour process to the 'formal' control already involved in the employment contract. On the other side of the coin, the task of the worker, which once involved an integration of conception and execution, has been progressively reduced to mere obedience to managerial instructions in which the conceptual phase of production is now incorporated.

In the interests of further cheapening labour power and of rendering it even more transparent to managerial control, the separation of conception and execution also makes it possible to redesign, fragment and simplify the labour of production. Braverman argued that, in this sense, there had occurred a 'degradation' and 'deskilling' of work over the course of the twentieth century in capitalist economies.

The object of this chapter is to examine the adequacy of some recent empirically-based claims to have refuted Braverman on the specific question of new technology. In the light of the foregoing, the first point to be made is that Braverman's interpretation of technological change is only a part – and arguably a subordinate part – of his overall thesis, most of which remains untouched by the attentions of empirically minded sociologists. Despite the numerous criticisms which have been made of Braverman (see Elger, 1982 and Salaman, 1982 for examples), no one now seriously doubts that scientific management was about the control of the labour process and that it involved the deskilling and degradation of work. Equally, no one outside the giddy world of management consultancy disputes that it remains 'the bedrock of modern work design' (Braverman, 1987). The following, for example, appeared in a 1970 publication by the Institute of Works Managers (Gilchrist, 1970: 123) to which practical managers contributed on the basis of their own experience, in this case, of ICI:

> The pioneering work of F. W. Taylor in the field of production efficiency led him to develop his plan of functional foremanship whereby the mental and clerical aspects of production were removed from the shopfloor and placed in the hands of functional specialists in the office. While Taylor's plan as a whole is subject to some fundamental weaknesses, the process of transferring the planning and control of production from the shopfloor to the office has continued and is now a virtually universal practice.

Even that part of Braverman's thesis which is concerned with technical change is only partially engaged by empirical work which focuses exclusively upon contemporary microelectronics. On the question of the effect of technological developments in textiles early in the capitalist era, for example, the evidence is substantially with Braverman. It is well established that the motive behind several key innovations in the textile industry was *precisely* to eliminate the skill monopolies of troublesome groups of workers (Marx, 1976: 563; Bruland, 1982), and that despite the 'worker resistance' which Braverman is so often accused of neglecting, they had exactly this effect. Note that studies of this type do not claim that *all* technological change was of this character, nor do they deny that new skills were created even as old ones were destroyed. What they *do* demonstrate, however, is the existence of a deskilling dynamic – or 'law of motion' – intimately linked with the operation of the capitalist economy.

The narrow focus of 'tests' of Braverman which concern themselves solely with new technology throws upon them a considerable burden of proof. Assuming, for the moment, that such studies succeed in proving Braverman wrong on a particular application of new technology, they can reasonably be asked why capitalists should behave differently when microelectronics rather than the managerial organization of work is the instrument involved, and why contemporary employers should use microprocessors differently from their forefathers' use of machinery.

Nor, finally, *is* it the case that even contemporary studies of microelectronics are unanimous in contradicting Braverman. For example, Noble (1977, 1984) found that American employers adopted numerical control (NC) machine tools in preference to the then more efficient record/playback technology because the former reduced their reliance upon skilled operators. Thompson and Bannon (1985) report that deskilling was involved in Plessey's introduction of their new 'System X' telephone switchgear. Even Davies' recent study of the brewing industry, determinedly optimistic as it is on the prospects for trade union influence, contains much evidence supportive of Braverman. As for white-collar workers, Cooley (1981), on the basis of his experience as a design engineer at Lucas Aerospace, described the 'Taylorization of intellectual work' as a result of the introduction of computer-aided design and Crompton and Jones (1984: 53) concluded that computerization had deskilled clerical work – more especially, female clerical work – in banking and life insurance firms. In the face of this, one might expect that empirical work on microelectronics which finds against the Braverman thesis should offer some explanation of why their findings should differ from those cited above: yet this is rarely done. Instead, the usual practice has been simply to dump the findings into a scale-pan marked 'anti-Braverman' and leave it at that.

Of course the requirement to engage opposing views cuts both ways. More specifically, those who believe that the Braverman thesis is broadly correct can reasonably be asked to account for findings which appear to contradict it. That, in a sense, is what this chapter is about. More specifically, it is concerned with the level of understanding of Braverman's arguments exhibited in some recent investigations of microelectronic applications. It is perfectly reasonable to be more interested in micros than in Braverman, but if the findings of empirical work are to be used to comment on his writings, these should at least be adequately understood.

The 'Law' of Deskilling
In their empirically-based critiques, Penn and Scattergood (1985: 613), Jones (1982: 197–9) and Wilkinson (1983: 14–15) credit Braverman with some invariant 'law' of deskilling as a consequence of capitalist development. If true, this would, strictly speaking, render his thesis vulnerable to disproof by the production of a single counter-instance – though, to their credit, none of the authors cited pushes the argument to this limit. This, however, is what Braverman actually says about the matter:

To the next question – how is the labor process transformed by the scientific–technical revolution – no such unitary answer may be given. This is because the scientific and managerial attack upon the labor process over the past century involves all its aspects; labor power, the materials of labor and the products of labor (1974: 169).

In other words the effects upon skills of individual instances of technological innovation may vary because innovation may be aimed at new products and processes rather than the labour process itself. Moreover, even when control and deskilling *are* on the agenda, the outcome in individual cases can certainly not be predicted on the basis of some inflexible 'law':

> This displacement of labor as the subjective element of the process and its subordination as an objective element in a productive process now conducted by management is an ideal realized by capital only within definite limits and unevenly amongst industries. The principle is itself restrained in its application by the nature of the various specific and determinate processes of production. Moreover its very application brings into being new crafts and skills and technical specialities which are at first the province of labor rather than management. Thus in industry all forms of labor coexist: the craft, the hand or machine detailed worker, the automatic machine or flow process (1974: 172).

This could scarcely be more lucid, yet misunderstanding is so widespread that a paraphrase is perhaps in order. Braverman in this passage made it clear that the separation of conception and execution in the productive process (which is the greater part of what he means by deskilling), far from being an invariant law, depends upon the opportunities presented by particular labour processes. Further, it may be accompanied by the temporary creation, within the workforce, of new skills. Finally, because of this last tendency, deskilling is never fully accomplished and thus cannot be understood as a progressive and simultaneous degradation of all labour processes.

It is clear, then, that the production of instances in which technological changes have not been accompanied by deskilling or have actually resulted in the creation of new skills is not sufficient in itself to contradict, let alone demolish Braverman's argument. However inconvenient it may be for empirical research, the fact is that Braverman – like Marx – was attempting to propound a 'law of motion' of capitalist society. If the legitimacy of such an enterprise is accepted at all, it can only be engaged by empirical work at its own level. Unexplicated counter-instances no more disprove such laws than rockets or hurled stones disprove the law of gravity. Nor should this be construed as an attempt to remove Braverman altogether from the reach of empirical enquiry. What *would* disprove Braverman would be evidence that skill levels within the working class as a whole were static or increasing over some reasonable period. In this respect More's (1982) argument that apprenticeship survived in the British engineering industry, not because of trade union pressure, but on the basis of the employers' continued reliance upon real craft skills comes closest to engaging Braverman as he should be engaged. That particular study, however, is vulnerable to the charge that it was centred on an industry which was particularly stagnant in both technical and managerial terms so that the deskilling dynamic, which is grounded in competitive capital accumulation, could have been in abeyance.

The 'Full Circuit of Capital'

Despite Kelly's (1985) claim to have 'moved beyond' Braverman in considering the results of management action throughout the 'full circuit of capital', it is evident from the foregoing quotations that Braverman *was* aware that the search for new products and processes interrupts and sets limits on the parallel search for means of deskilling and extending management control of labour power. Since he explicitly states that the outcome, locally and for a time, can be a reskilling of the labour process, this means that Braverman does *not* propound a universal law of deskilling. What he *does* claim is that there exists a general tendency for deskilling to occur in capitalist economies which will become actual where products and processes make this possible and where its effects are not masked by initiatives aimed at changing technology for other reasons.

In the case of the introduction of NC machine tools by the aerospace contractors studied by Jones (1982: 191), it is apparent, on the author's own account, that an important management motive was the desire to offer products not even possible with conventional machine tools. Why, then, should this (not very convincing) evidence that there was no deskilling be regarded as inconsistent with Braverman? Beyond telling us that the industry faced intense foreign competition, Penn and Scattergood (1985) reveal little of the motives which led to computerization by their cardboard manufacturers. However they do tell us that, in two of the firms, computerization was followed by large-scale redundancies amongst warehouse staff and that productivity in all of the firms roughly doubled. Why should it be held to contradict Braverman that, whilst the management were achieving these gains, there was no deskilling of production work and even the creation of new skills amongst maintenance workers?

Braverman as a Technological Determinist

In the conclusion of his empirically-based rejection of the deskilling thesis, Jones (1982: 198) states that 'there is nothing "inherent" in the hardware of NC or its concept that would allow for the deskilling and control and surveillance assumed by both theorists of the labour process and publicists for NC installation.' If Braverman was the leading theorist of the labour process, this can only mean that he stands accused of technological determinism. Albeit in a piece more concerned with Marx's than Braverman's arguments on deskilling, precisely the same accusation is made by Penn (1983: 32).

This is what Braverman actually said in his discussion of NC machining, which is exactly what Jones' study is concerned with:

> there is no question that from a practical standpoint, there is nothing to prevent the machining process under numerical control from remaining the province of the total craftsman. That this almost never happens is due, of course, to the opportunities the process offers for the destruction of craft and the cheapening of the resulting pieces of labor into which it is broken (1974: 199).

Since this makes precisely Jones' point, one is left wondering who are these 'theorists of the labour process' who make such rash pronouncements about the 'inherent properties' of NC machine tools. One might also add that, in two of the firms which Jones studied, NC machines were being worked by the bottom grade of operatives, under the direction of a setter, after just one month's training. This is exactly what Braverman's quotation would lead us to expect, *not* on the basis of some inherent property of NC machining but because, in a capitalist enterprise, the opportunities which it presents for deskilling are likely to be seized upon.

Of course there are writers who maintain that there exists a 'capitalist technology' which, as the product of capitalist social relations of production, has become incompatible with other forms of work organization (e.g. Gorz, 1976: 160–1). Such global versions of technological determinism are an easy target but are absent, so far as I am aware, from writings in the labour process tradition. On the other hand it would be a form of bigotry to dismiss out of hand the idea that some technologies, once having been installed, *do* exert at least a constraining influence on the social relations of production. Thus when Cockburn (1983: 121) remarks that information technology is precisely *about* control from above, the issue is at least worth discussing. And when Rosenbrock (1977) called for computer technology which would enhance rather than destroy skills, he did not suppose that this could be achieved without altering the technology itself.

Although no technology, in the ultimate, determines the use which will be made of it, the theoretical inadequacy of technological determinism does not mean that the implications of technical change for the labour process are purely a function of the negotiations (in the widest sense) surrounding its introduction. Davies' (1986) study can be read as a sobering illustration of the limits on what trade unions can achieve after the decision to introduce new technology has been taken. To use an example from a research in which I was involved some years ago: if a chemical plant is designed with a control room several hundred feet up in the air, someone is going to have to work in it. Indeed it is precisely in this sense that manual workers can be said to be 'living thinkwork' previously performed by intellectual labour working in the service of capital (Hales, 1980).

The Meaning of Deskilling in Braverman

For Braverman the definition of skill starts with the unity of conception and execution in traditional craft production. Under capitalist social relations of production, this unity has been progressively destroyed and replaced by a division of labour in which scientific and technical knowledge is concentrated within managerial and staff organizations. It is this loss of mastery of the conceptual aspect of production which is at the heart of deskilling in Braverman's usage of the term, *not*, in the first instance, the fragmentation or simplification of the task; although these tend to follow as consequences and Braverman, of course, devotes much attention to them. Of recent empirical studies of the labour process, that of Crompton and Jones (1984: 59) stands out in its clear recognition of this fundamental control aspect of deskilling.

Though he has been accused many times of romanticizing a past of craft production which perhaps never existed, Braverman's way forward is not a return to some pre-scientific arcadia but a form of collective organization in which 'the worker can regain mastery over collective and socialized production only by assuming the scientific, design and operational prerogatives of modern engineering; short of this there is no mastery of the labor process' (1974: 445). In Braverman's own words, he is motivated not by a nostalgia for the past but 'by an age which has not yet come into being' (1974: 7).

The same passage also contains what might be read as Braverman's anticipatory answer to arguments centred on the idea that reskilling can come about as a result of management training or as a result of the development of 'tacit skills' gained by simple experience or repetition (Manwaring and Wood, 1985). Given the managerial appropriation of the scientific and technical knowledge necessary to unify the labour process:

> What is left to workers is a reinterpreted and woefully inadequate conception of skill: a specific dexterity, a limited and repetitious operation, speed as skill. . . . The instruction of the worker in the simple requirements of capital: here in the minds of managers is the secret of the upgrading of skills so celebrated in the annals of modern industrial sociology (1974: 443).

The important moral here is that so long as management-devised training for newly 'reskilled' tasks excludes the scientific or technical knowledge necessary to unify the productive process, it is not, in Braverman's terms, reskilling at all since the power to make all meaningful decisions on production remains concentrated in management hands. The worker is merely being retrained to the point where she can carry out revised operations in a system in which conception and execution remain as divided as ever. Again, if skill is seen as the ability to comprehend, plan and execute a total process of production, it is not really separable into discrete abilities as is implied by current training jargon ('skills'), nor can there be a meaningful replacement of operational by 'social skills' as is sometimes argued since these, of themselves, produce nothing.

The importance of this aspect of deskilling and the manner in which it opens up the labour process to capitalist controls is clearly brought out in Fairclough's (1986) study of the degeneration and demise of the Meriden Motorcycle Co-operative. Faced with technical problems on the Bonneville motorcycle, the Meriden workers, though 'skilled' craftsmen as the term is commonly used, were forced to depend for help on engineers from GEC. However the engineers imported with them a conception of authority relations normal in the capitalist social relations of production but alien to the democratic ideals of the co-op. Their presence thus helped to reinforce other degenerative tendencies. The point is that it is an inability on the part of the workforce to cope with the totality of the productive process which is the key to deskilling. It is this which opens up the labour process to capitalist control via the intellectual component of productive labour – not so much the absolute simplification of the task itself.

Indeed, this leads to a line of thought which was only briefly explored by Braverman himself (1974: 425): that deskilling should really be thought of as *relative* to the totality of the knowledge and skills necessary to comprehend the productive process. In this sense, the application of science to the productive process, if wholly incorporated within the structures of capitalist management, could be said to result in a relative deskilling of (for example) tradition-based machine-shop crafts, without any physical changes occurring in the latter at all.

Equally, it is quite possible that relative deskilling and the kind of absolute task simplification characteristic of Taylorism could occur at the same time. For the most part, of course, Braverman's thesis concentrates upon the latter aspect of deskilling.

Reskilling and the Problem of Levels

When discussing the issue of 'reskilling' it is important to be precise about where, in capitalist work organization, skills are created and destroyed. Thus Crompton and Jones (1984: 74) note that, whilst the overall effect of computerization in banks and life insurance companies has been to deskill clerical work, a small number of skilled jobs have been created for programmers and systems analysts. These, however, are not accessible to clerical workers through the normal channels of training and promotion and the overall result has been to stratify the workforce all the more rigidly. Though true, this slightly misses the point. The work of programmers and, more especially, systems analysts, is concerned primarily with devising systems of work control and is to that extent a managerial task. So far from providing counter-instances to a general deskilling tendency, what amounts to a further concentration of knowledge and skill in the managerial strata is precisely what the deskilling thesis is about. Jones (1982: 200) is clearly aware of this issue, though he raises it only to dismiss the idea that the 'control' function of part-programming NC machine tools can be divorced from craft skills and thus monopolized within managerial strata. In fact there is every possibility that this could happen since it is characteristic of British lower and middle management that they tend to be recruited directly from the shopfloor (e.g. Granick, 1971: 56).

The Social Definition of Skill and Trade Union Action

Though the ultimate test, for Braverman, of whether or not workers are skilled is what they can actually *do*, first individually and latterly collectively, he was certainly not unaware that there is a question of the social definition of skill. Though he does not discuss the maintenance of such a definition by trade union power after its substantive basis has disappeared – a process which Penn (1982: 104–5) considers to disprove the deskilling thesis – he has quite a lot to say about the redefinition of skill by employers and academic commentators:

> With the development of the capitalist mode of production, the very concept of skill becomes degraded . . . with jobs requiring quite short periods of training being classified as 'skilled' (1974: 444).

Perhaps luckily, Braverman did not live to see some commentators claiming 'tacit skills' on behalf of jobs requiring no training at all (Manwaring and Wood, 1985).[1]

For Braverman such social redefinitions of skill were evidently to be exposed and challenged, not taken uncritically as measures of whether or not deskilling has occurred. Certainly the statement of Penn and Scattergood (1985: 617) that the job of the 'beatermen' responsible for overseeing the beating of pulp to make cardboard 'was regarded traditionally by management and workers as a highly skilled occupation requiring at least four years' learning as a beaterman's assistant' is the kind of social definition of skill that Braverman would not have accepted at face value, particularly since we are also told that the beatermen's 'knowledge of what to do in any given circumstance is a matter of judgement based primarily on experience'. As an AUEW shop steward explained some years ago to the author, 'there's a difference between skilled and experienced'. Though the Penn and Scattergood study claims to contradict Braverman in that the computerization of the production process did not deskill the work of these 'beatermen', a more reasonable interpretation is that they were unskilled both before *and* after the technical changes, as indeed is stated in the authors' own introduction (1974: 611). In this sense, the research has little to say about the deskilling thesis one way or the other, particularly since the continued attribution of skill was also said to have depended on trade union organization, a theme which recurs from Penn's earlier work (1982).

If it is illegitimate to compare the vagaries of 'social definitions' of skill with what Braverman says about the actual fate of skilled work in capitalist societies, it is still more so to credit him with the kind of subjectivist industrial sociology which has spawned a thousand surveys of 'work satisfaction', 'worker alienation' and the like. Thus, in the face of Braverman's declared intention to investigate what was happening to the working class as a class *in itself* (1974: 27), it is a little surprising to read that 'the deskilling theory as elaborated by Braverman is about the effects of deskilling on workers' consciousness' (Penn and Scattergood, 1985: 620) and even more so to be told that Braverman's thesis is contradicted by instances in which workers do not *think* they have been deskilled.

Braverman is also taken to task by Cockburn in her interesting discussion of the issues involved in deskilling (1983: 113ff.), for a failure to distinguish between the skill involved in a *task* and the skill possessed by the *worker*, pointing out that the skills possessed by apprentice-trained compositors in the newspaper industry are now scarcely relevant to the jobs now performed by them, nor to the defence by them of a gender-based skilled status through trade union organization. In fact Braverman was aware of the former distinction as is shown by his discussion of the growing disjunction between educational requirements and the routinized reality of most industrial work (1974: 436ff.). On the later point, it is perfectly true that his concentration upon the pressures to which the working class has been subjected rather than upon its response also means that he does not discuss the maintenance of skilled status in the face of a deskilled reality by means of craft union organization. Perhaps the real answer to Cockburn's point has been delivered by Rupert Murdoch – that the artificial maintenance of skilled status is ultimately a transitional phenomenon which Braverman was entitled to neglect when concerned with developments over a century of capitalist development and across the entire economy.

The Transitional Politics of Technical Change

Just as the society-wide scope of Braverman's conception creates problems for plant-level case studies which set out to test his thesis, so its historical sweep is difficult to relate to the time-scale of practical fieldwork. Almost inevitably, case studies are normally limited to the transitional politics surrounding the actual introduction of technical change in particular workplaces. This is explicitly so in the studies reported by Wilkinson (1983) who states that his empirical material 'focuses mainly on the debugging and implementation stage of innovation'. Convincingly Wilkinson demonstrates that these stages provide 'junctures' at which workers can seriously attempt to impose their own interests, and sometimes succeed, despite the fact that in at least two of his four cases, an important motive for introducing the new technologies was an explicit management desire to wrest control from the shopfloor. However, these short-range successes on the part of the workforce do not really contradict Braverman as Wilkinson appears to believe. For that one would have to establish, if not by direct evidence, at least by reasoned argument, that the workforce has a reasonable prospect of maintaining working practices established by them and acceded to by management during those initial periods of crisis and alarm when managements have every interest in the continuing presence of skilled workers able to pick up the pieces whenever newly introduced processes go wrong. In arguing that these manning practices will survive the introductory phases of new technologies, we are back once more with the question of 'artificial' working practices imposed by union power alone and lacking a substantive base.

In fact there is, in the industrial relations literature, some basis for arguing that 'custom and practice' constitutes such a mechanism (Brown, 1972), but Armstrong and Goodman (1979) have shown that this mechanism is equally available to managements when the balance of power shifts in their favour. Moreover the long-term fragility of working practices maintained by union organization alone has been demonstrated by the marked increase of working flexibilities during the current recession (Edwards, 1985a,b).

The studies reported by Jones (1982) and Penn and Scattergood (1985) also clearly report transitional effects in that skilled labour was retained on routine tasks, in both cases primarily in order to cope with breakdowns or to make good temporary design inadequacies in the new processes. It scarcely follows that the long-term effects of the new technologies on skilled employment will be similarly minimal. At one time, after all, the spread of the automobile created quite favourable employment prospects for men with red flags.

Automation Hype and the Thought Processes of Managers

It is not difficult to discover deskilling motives within capitalist managements. For example Davies (1986: 131) quotes a brochure for automated equipment in the brewing industry which claims

> to replace the intelligent functions of man by instrumentation and engineering. The brain is replaced by the control unit, the eye by the sensing device and the hand by the operating device.

The fact that this is automation hype by a huckster of new technology matters not a jot. Such persons and their motives are just as much a part of the capitalist management process as the salaried functionary. However, not everyone involved in the management process says such things and some would even flatly contradict them. The question then is, how does this diversity of motive reflect upon the Braverman thesis?

It is a long-standing criticism of Braverman that he treats 'capital' in an undifferentiated manner (e.g. Salaman, 1982: 58–9) and that in his schema 'capital' appears to act in a unitary fashion so as to extend 'its' control of the labour process, with little regard paid to the motives or tactics adopted by managers, still less to the fact that these might vary even within individual workplaces. If one is interested in the *differences* in the way in which managers or managements act, or in the experiences of managers or in a variety of similar questions, such criticisms are quite valid. However, it is clear that Braverman himself was interested in general system-wide tendencies and only in the impact of these upon the working class. At this level of analysis, it is arguable that too great an emphasis on the motives of managers would present a misleadingly voluntaristic picture of what is going on.

Capitalist management is, in its contemporary version, a highly differentiated system. It follows that the imperative of accumulating surplus value will appear in different guises at different points in the system – and so will subsidiary imperatives which derive from it. Thus one management may introduce new technology 'because of the competition' (Penn and Scattergood, 1985: 623), whilst individual managers in other cases might mention reductions in 'floor-to-floor' time, the scarcity of skilled labour and so on (Jones, 1982: 191). Even taken at face value, this diversity of motive is not evidence that deskilling or the extension of management control plays no part in the introduction of a new technology. In a complex organization, it is not necessary for each manager to make decisions with the explicit goal of extending control of the labour process. It is enough for them to pursue their own parts in the differentiated process by which this is achieved.

In fact even this degree of purposiveness may be unnecessary. The ability of organizations to harness good intentions to indifferent ends means that new technologies originally introduced, say, to increase safety or to offer new services can open up deskilling possibilities which, although no part of the original management intentions, may then be activated later and by a different set of managers. Cases in point are the effects of computerization and on-line data systems on Post Office clerical workers (Batstone *et al.*, 1984: 154) and some of the effects upon signalmen and traffic controllers of the automatic signalling equipment introduced by British Rail (Pendleton, 1986: 267–84). The fact – where it is a fact – that there may have been no explicit deskilling motive behind the introduction of such equipment (*ibid*) means very little as far as the workforce is concerned. Certainly it does not contradict Braverman.

Management organization can even accommodate motives diametrically opposed to deskilling. The work of Wilkinson (1983) testifies to the variety of self-interested and other motives held by managers involved in the politics of decision-making on new technologies, some of which can amount to managerial collaboration with the resistance of workers to its introduction. From the point of view of the enterprise as a whole, it is enough that those motives compatible with the extension of control eventually prevail.

All of this assumes that managers' motives are what they say, or think, they are. It is not so much that managers might not tell the truth to sociologists (Jones, 1982: 191); rather that managers' motives are likely to include *tacit* assumptions concerning the labour process. Thus the influential management writers Hedberg and Mumford (1975) have pointed out the 'woefully limited model of man' tacitly assumed by the designers of new technologies and systems, assumptions of which they are, themselves, largely unaware of (see also Thompson, 1986). In other words, it is arguable that a deskilling dynamic is now implicit in contemporary conceptions of good engineering or systems analysis practice in capitalist societies.

In summary, if management motives are to be the means of demolishing Braverman, it is necessary that:

1. tacit motives are properly allowed for;
2. the motives in question, when mediated through the organizational position of those who hold them, are actually incompatible with deskilling or the extension of control;
3. such motives dominate the decision-making process;
4. the long-term survival of the enterprise is not prejudiced by the resultant decision.

If this appears to be a tall order, so, one might suggest, is the refutation of Braverman's thesis.

The Wider Impact of Technological Innovation

Empirical work on new technology persistently treats job losses and deskilling as separate categories (see, for example, Davies, 1986) in a way that would be impossible if the studies were concerned, as Braverman was, with the progress of the working class as a whole. Needless to say, such studies never follow up the stories of those displaced and, in consequence, we lack data on the impact of redundancies on the overall levels of skill within the working class. One would guess that the deskilling impact is considerable.

Even within the employed population, there are 'knock-on' effects of technological change to be considered once it is recalled that managerial work reorganization is always on the agenda as a possible response to intensified competition. In other words, technological changes which do not, in themselves, result in deskilling, may nevertheless trigger off a competitive search for means of labour intensification (including deskilling) in areas where the technical changes in question have *not* occurred.

An instance of this was reported by Thompson and Bannon (1985: 79). The Plessey management, having begun production of electronic switchgear at their Edge Lane plant found that the existence of the new processes gave them 'massive leverage to attack the working practices, skills and pay structures used by workers to their advantage' in areas of traditional production. In the short run, of course, the precise outcome of such 'knock-on' tendencies depends on the balance of power; Penn and Scattergood (1985) report that 'the co-existence of traditional skilled work and newer technologies in the same plant acts as a restraint on management' but this obviously cannot be regarded as a general rule.

For empirical studies of deskilling, the implication of these obser-vations is that the overall impact of technological change upon skills cannot be assessed so long as the case study material is confined only to those employed – and who continue to be employed – in those plants where innovation actually takes place. For example, if productivity roughly doubled in the five computerized cardboard-making plants studied by Penn and Scattergood (1985), without much local effect upon skill levels, what, one is left wondering, were the effects in firms which did *not* computerize?

Conclusions

Despite an evident desire in some quarters to refute Braverman's analysis on the question of new technology or to 'move beyond' it in some way, it is questionable whether either has been achieved to any significant extent. Any sensitive reading of his work should reveal that Braverman actually regarded the deskilling tendencies of technical change as a system-wide dynamic or 'law of motion' in capitalist economies which could, temporarily and locally, be interrupted or reversed by a variety of factors, many of which have been rediscovered by his critics as supposed refutations. In particular, some of the case studies which have set out to 'test Braverman' suffer from the perhaps inherent defect that they have actually been concerned with the developmental stages following the introduction of new technology rather than with the long-run effects and with local results rather than with system-wide repercussions. Additionally, however, certain studies have unwittingly redefined Braverman's conception of skill by omitting the crucial 'control' element. Some have confused the 'feeling' of skill or the retention of skilled status by trade union action with the actual substance of skill. Moreover, in some studies, the absence of an explicit deskilling motive amongst managers involved in technological innovation has mistakenly been held to contradict Braverman. Finally, as if all this were not enough, the 'Braverman' who has been refuted has sometimes been a crude social or technological determinist, which the real Braverman certainly was not.

Note

1. Of course where the denial of formal training is a device used by employers to avoid paying skilled rates, the proponents of the 'tacit skills' thesis have a point. For women trade unionists, however, there is a pitfall in the argument that simple jobs are 'really' skilled. Whilst the intention may be to secure equality of treatment with male jobs which have been defined as skilled by exclusionary trade unionism, the effect may also be to obscure

the real processes of deskilling as they impact disproportionately upon female employment.

References

Armstrong, P., and J. D. Goodman. 1979. 'Managerial and Supervisory Custom and Practice'. *Industrial Relations Journal* Vol. 10, no. 3, 12–24.

Batstone, E., A. Ferner and M. Terry. 1984. *Consent and Efficiency*. Oxford: Basil Blackwell.

Braverman, H. 1974. *Labor and Monopoly Capital: the Degradation of Work in the Twentieth Century*. New York: Monthly Review Press.

Brown, W. 1972. 'A Consideration of Custom and Practice'. *British Journal of Industrial Relations*. Vol. 10, 42–61.

Bruland, T. 1982. 'Industrial Conflict as a Source of Technical Innovation: Three Cases'. *Economy and Society*. Vol. 11, no. 2, 91–121.

Cockburn, C. 1983. *Brothers: Male Dominance and Technological Change*. London: Pluto.

Cooley, M. 1981. 'The Taylorisation of Intellectual Work'. *Science, Technology and the Labour Process: Marxist Studies Vol. 1*. Ed. C. Levidow and R. Young. CSE Books, 46–65.

Crompton, R. and G. Jones. 1984. *White Collar Proletariat: Deskilling and Gender in Clerical Work*. London: Macmillan.

Davies, A. 1986. *Industrial Relations and New Technology*. London: Croom Helm.

Edwards, P. K. 1985a. 'Managing Labour Relations through the Recession'. *Employee Relations.* Vol. 7, no. 2, 3–7.

Edwards, P. K. 1985b. 'Managing through the Recession: the Plant and Company'. *Employee Relations*. Vol. 7, no. 3, 4–8.

Elger, A. 1982. Braverman, Capital Accumulation and Deskilling. *The Degradation of Work? Skill, Deskilling and the Labour Process*. Ed. S. Wood. London: Hutchinson, 23–57

Fairclough, M. 1986. 'The Political Economy of Producer Co-Operatives'. Unpublished Doctoral Thesis. University of Bristol.

Gilchrist, R. R. 1970. (ed.) *Works Management in Practice* London: Heinemann (for the Institute of Works Managers).

Gorz, A. 1976. 'Technology, Technicians and Class Structure'. *The Division of Labour*. Ed. A. Gorz. Brighton: Harvester Press, 159–89.

Granick, D. 1971. *Managerial Comparisons of Four Developed Countries: France, Britain, The United States and Russia*. Cambridge, Mass.: MIT Press.

Hales, M. 1980. *Living Thinkwork: Where Do Labour Processes Come From?* London: CSE Books.

Hedberg, B., and M. Mumford. 1975. 'The Design of Computer Systems: Man's Vision of Man as an Integral Part of the System Design Process'. *Human Choice and Computers*. Ed. E. Mumford and H. Sackman. Amsterdam: North-Holland, 31–59.

Jones, B. 1982. 'Destruction or Redistribution of Engineering Skills? The Case of Numerical Control'. *The Degradation of Work? Skill, Deskilling and the*

Labour Process. Ed. S. Wood. London: Hutchinson, 179–200.

Kelly, J. 1985. 'Management's Redesign of Work: Labour Process, Labour Markets and Product Markets'. *Job Redesign: Critical Perspectives on the Labour Process*. Ed. D. Knight, H. Willmott, and D. Collinson. Aldershot: Gower, 30–51.

Manwaring, T., and S. Wood. 1985. 'The Ghost in the Labour Process'. *Job Redesign*. Ed. D. Knight, H. Willmott, and D. Collinson. Aldershot: Gower, 171–96.

Marx, K. 1976. *Capital Vol. 1*. Harmondsworth: Penguin.

More, C. 1982. 'Skill and the Survival of Apprenticeship'. *The Degradation of Work? Skill, Deskilling and the Labour Process*. Ed. S. Wood. London: Hutchinson, 109–21.

Noble, D. F. 1977. *America by Design*. Oxford: Oxford University Press.

Noble, D. F. 1984. *Forces of Production*. New York: Alfred Knopf.

Pendleton, A. 1986. 'Management Strategy and Labour Relations in British Rail'. Unpublished Doctoral Thesis. University of Bath.

Penn, R. 1982. 'Skilled Manual Workers in the Labour Process 1856–1964'. *The Degradation of Work? Skill, Deskilling and the Labour Process*. Ed. S. Wood. London: Hutchinson, 90–108.

Penn, R. D. 1983. 'Theories of Skill and Class Structure'. *Sociological Review*. Vol. 31, no. 1, 22–38.

Penn, R. and H. Scattergood. 1985. 'Deskilling or Enskilling? an Empirical Investigation of Recent Theories of the Labour Process'. *British Journal of Sociology*. Vol. 36, no. 4, 611–30.

Rosenbrock, H. 1977. 'The Future of Control'. *Automatica*. Vol. 13, 389–92.

Salaman, G. 1982. 'Managing the Frontier of Control'. *Social Class and the Division of Labour: essays in honour of Ilya Neustadt*. Ed. A. Giddens and G. Mackenzie. Cambridge: Cambridge University Press, 46–62.

Thompson, P. 1986. 'Crawling From the Wreckage: the Labour Process and the Politics of Production'. Paper presented at the Fourth Annual Labour Process Conference, Birmingham: Aston University.

Thompson, P., and E. Bannon. 1985. *Working the System: The Shopfloor and the New Technology*. London: Pluto.

Wilkinson, B. 1983. *The Shopfloor Politics of the New Technology*. London: Heinemann Educational.

9

Skills, Options and Unions: United and Strong or Divided and Weak?

Jon Gulowsen

In some workplaces all or most workers are organized in trade unions. By acting collectively, they are able to obtain proper wages and good working conditions. In other places the unions have little support and the workers either put up with poor conditions or leave the company, hoping to find better conditions elsewhere. Thus there exist large differences among the tactical options available to workers. How are these to be explained? What determines the options, the strategic position, of the workers in a company?

A number of answers are offered: class consciousness, varying support from national unions, the size of the workforce etc. In this chapter I will focus on skills: How do these affect the strategic position of the workers in different workplaces? And how does the development of skills reflect the interests of workers and managers?

Such questions are encouraged by recent initiatives in work reorganization under the banner of 'quality of working life' (QWL). Emery (1959) and Emery and Thorsrud (1976) suggest alternatives: job enlargement, job rotation and partly autonomous work groups, principles that imply increased or reorganized skills in the workplace. In Norway these principles were originally launched as an attempt to enhance industrial democracy; but although they often have positive effects on worker satisfaction and productivity, their implications for power relations are more dubious.

The Theoretical Background

According to Marx, capitalist society tends to develop into two antagonistic classes as the forces of production develop. Marx has no explicit concept of power. Every aspect of society, the state, culture and religion, as well as the conditions in industry, reflects the powerlessness of the working class. Carrying on where Marx left off, Braverman (1974) deals with the development of skills under capitalism. In his book *Labor and Monopoly Capital* with the illustrative subtitle: 'The degradation of work in the twentieth century', he argues that workers' skills have generally deteriorated as a result of capitalist development. Though Braverman studies the relationship between skills and power relations in the workplace, his focus is different from mine. Like Marx he is concerned with the relationship between classes, his method and concepts are not developed in order to study differences within the working class. On the contrary, he argues that the working class is gradually becoming more homogeneous.

According to Weber, ownership structure and corresponding market relations determine class relations. Pointing to a number of conditions that affect market position, including skills, Weber argues that capitalist society includes a variety of classes. Addressing Marx he lists a number of reasons why organized class behaviour is unlikely to develop.

The approach of this chapter is Weberian. By focusing on the strategic position of the workers rather than class relations, it takes the actor's point of view. It tries to understand differences rather than explain structures and forecast trends.

Strategic Position

Faced with problems in the workplace, workers may respond in different ways. Some choose to put up with the problems and remain passive. Others act. They have two types of options: to leave the employer and look for better jobs in the labour market (exit), or to try to work out changes (voice) (Hirschman, 1970). Individual and collective strategies are available. Some try to improve their own situation, for example, by advancing within the company hierarchy. But most workplace problems cannot be solved by individual strategies. Another strategy is to try to improve conditions collectively by organizing trade unions. These concepts focus attention on three different arenas: the labour market, the company and the union. In some workplaces the workers have a variety of options; elsewhere they are more limited.

Skills

Historians and statisticians often divide workers into three categories: skilled, semiskilled and unskilled. While useful as indicators of the general level of education in a society, and of differences within the working class, the concepts are imprecise and say little about the potential of the workers. In an attempt to identify how skills affect strategic position, I find three dimensions useful. I will stick to the traditional distinction between higher and lower skills. Among higher skills I will distinguish between two additional dimensions:

1. *General versus company specific skills.* General skills are useful in many firms and include traditional skills acquired through apprenticeship training and vocational school systems and other skills of general relevance. Company specific skills are limited to one firm. The training is usually administered by the company training unit and focuses on the problems and procedures of that firm. Altmann and Bøhle (1972) have shown that workers with general skills have a better bargaining position and that this affects their ability to improve the work environment and prevent accidents. Company specific skills correspond to the term internal labour market (Doeringer and Piore, 1971).

2. *Social versus technological definition of skills.* More (1980) makes a distinction between social and technological definitions of skills corresponding to formal and genuine skills. This is a useful point of departure, but it needs some elaboration. The definition of skills may reflect the interests of workers or employers. If the workers are sufficiently strong, they frequently try to formalize skills. In order to promote job security and to defend and expand established rights, they often try to prevent technological innovations which would make their skills obsolete. Such attempts correspond to More's term: social definition of skills.

It is generally in the interest of management to be able to benefit from scientific and technological progress and to adapt flexibly to the tasks connected with new technology. Previous formal skills may be an obstacle in this respect, as management is not interested in paying for obsolete skills. What they want is flexible skills, or in More's terms: technologically defined skills.

With both forms of skill definition, present conditions are the product of previous struggles and will affect future developments. The relative strength and the interest of workers and management will in each case determine their actual position on this dimension. Many strong trade unions have sustained social definitions of skills and have avoided serious technological threats. This has consolidated their strength. By contrast, in other contexts strong management has successfully gained control of skills and personnel development. This has made it more difficult for workers to influence the application of technology and the definition of skills.

Combining these two dimensions, we can identify four categories of skilled workers, while unskilled workers make up a fifth group.

1. *Higher, socially defined, general skills.* Artisans such as carpenters

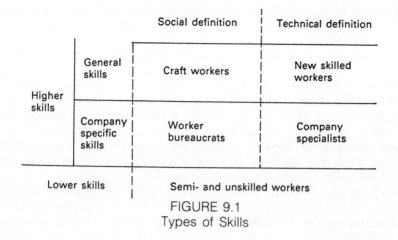

FIGURE 9.1
Types of Skills

and painters, and workers with traditional, industrial skills such as plumbers and electricians belong to this category. They may be termed craft workers.

2. *Higher, socially defined, company specific skills.* This category is predominant in public companies with a relatively stable, often bureaucratic task structure such as railways and post offices. They will be called worker bureaucrats.

3. *Higher, technologically defined, company specific skills.* Control room operators in technologically advanced, monopolistic industries are typical of this category which will be called company specialists.

4. *Higher, technologically defined, general skills.* Technicians in data processing and electronics, and makers of advanced instruments, belong to this group: the new skills.

5. *Unskilled workers.* This is a large and heterogeneous category. A focus on skills obviously does not catch the essential characteristics of the different groups of unskilled labour.

The Strategic Position of Different Categories of Skill

Craft Workers

Traditionally artisans and craft workers were trained by working with masters or experienced journeymen. This system has gradually been modified. At present vocational schools provide a significant part of the training in most industrial countries. However, both these systems involve a degree of standardization and formalization of skills, as well as norms of social behaviour. These standards are set by the crafts in response to the wide variety of demands from different companies.

This gives craft workers mobility in the labour market. From the point of view of the employers, however, much of what apprentices learn seems superfluous and frequently out of date.

Higher, general and socially defined skills provide a sheltered labour market for the craft worker. He can get a job wherever his skills are in demand. (The social definition of craft status is traditionally applied exclusively to male workers.) He does not have to compete with other categories of workers; he can improve his situation by leaving the company and find work elsewhere, usually without significant loss of wages or benefits. Thus he is relatively independent of the goodwill of any particular employer and can take part in trade union activity or stand up to the employer without fear of sanctions.

In principle, however, a craft worker is not irreplaceable. Standardized skills give the employer the opportunity to find replacements in the labour market. The ability to do so depends on the extent to which the relevant segment of the labour market is controlled by a trade union.

Craft workers have a unique position in the trade union movement. They organized earlier than most other groups of workers, and tradition and shared historical experiences support collective consciousness. Traditions of collective problem-solving and collective agreements encourage continued support for the union. Traditions also have practical implications. If a union has succeeded in gaining high wages and good working conditions, this motivates workers to join and support the union and tends to reduce labour turnover as well. If the employer recruits new workers, it will usually be easy for the union to gain their support.

While other categories of highly skilled workers, most notably the worker bureaucrats, may be ranked according to seniority or position within the firm, craft workers are in principle equals. This equality has two implications for union behaviour. Having similar rank and training, and being mostly paid according to the same principles, workers are likely to agree about union demands. All union members have the same moral right to take part in union activities and speak up on behalf of the trade union; there is little reason to yield to more experienced members.

Worker Bureaucrats

Worker bureaucrats mostly work in large organizations with stable tasks and highly bureaucratic structures, such as the national railways. Training schemes are adapted to the tasks of the organization and most vocational training is provided by company schools. Special courses correspond to different positions in the company hierarchy. Advancement is based on seniority among qualified applicants.

The value of the worker bureaucrats' skills is almost entirely limited to their company. Once they have advanced within the hierarchy and gained seniority and special benefits, leaving the company is difficult. All that is available outside is unskilled work. Thus when worker bureaucrats have grievances, they must settle them inside the firm. The bureaucracy gives them reasonably good opportunities to do so. One way is simply to advance out of the problems. With job security in a bureaucratic structure, they are likely to stand up directly to their superiors when they deem it necessary. They do not risk losing their jobs that way.

Worker bureaucrats have a strong basis for collective action. With almost negligible labour turnover, strong formal and informal networks are likely to develop. The union tends to make demands and choose means of action which are appropriate in a long-term perspective. Spontaneous and militant actions are avoided, as the strategic interests of the company often coincide with those of the employees. Unable to recruit qualified people from outside the firm, the company depends on having a stable workforce and tends to behave accordingly. Neither of the parties will push their own interests in a manner that might threaten the existence of the firm.

Certain characteristics of the career system tend to support this pattern. In a promotion system like the national railways, age and experience are prerequisites for advancement within both the company and the trade union. It is difficult for young, inexperienced workers to speak out on behalf of the majority and to run for office in promotion unions (Clegg, 1976). Shop stewards tend to be elected from the most experienced workers. As a result, this kind of trade union is apt to represent the interests of workers with high seniority rather than the bottom layers of the hierarchy. Stability and job security will be guidelines for such a union. Furthermore, it is in the interest of the union to avoid competition for advancement among its membership. This calls for unambiguous criteria of advancement. It follows that the union will support a bureaucratic development of the firm and resist attempts by management to weaken new bureaucratic routines in response to technology.

Company Specialists

Many fields with rapidly changing technology are dominated by one or a small number of firms. In such fields it is impossible for firms to recruit qualified workers in the open labour market. They have to train key personnel themselves. The normal procedure is to select workers who have shown interest and capacity to learn and have proved their loyalty to the company. These are given special training either by the firm or by suppliers of new machinery and may receive salary increases or other benefits. Company specialists are likely to have diverse and challenging tasks and a high degree of work autonomy.

The skills of company specialists may be rendered obsolete by technological progress after a few years. This has important implications for both workers and employers. Since their old skills become devalued internally and are possibly worthless outside the firm, the company specialists have to accept retraining if it is offered them. The company has two options. If satisfied with a worker's competence and loyalty, they may offer retraining. If not, they may give another person the job and train that person from scratch. The latter alternative may not be much more expensive. Trade unions are usually opposed, but unions in this field tend to be weak or even non-existent.

As with worker bureaucrats, there is a very strong mutual interdependence between company specialists and firms. Neither can find alternatives in the open labour market, but the relationship is asymmetrical. Though the skills of the company specialists are very valuable for the firm, they are usually even more valuable for the specialists themselves. They have a lot to lose from protesting. Having no tradition or skills in common, and being handpicked by management for their abilities and company loyalty, this group has little basis or incentive to unionize.

The following example illustrates the strategic position of company specialists (Gulowsen, 1974). Norsk Hydro is the sole manufacturer of chemical fertilizer in Norway. Prior to 1967 the fertilizer factory was manned by some 10 unskilled workers on each of four shifts. Each shift was headed by a foreman and an assistant foreman. A maintenance crew of skilled mechanics and plumbers and some unskilled labourers also worked the day shift. The four shifts were organized in the traditional scientific management manner. The process was divided into different sections: each of the unskilled process operators was responsible for one section, and very few were acquainted with other areas of the production system.

From the management's point of view this pattern of organization had important weaknesses. It was rigid and poorly suited to coping with production problems. There were significant direct and indirect costs associated with a high rate of absenteeism and labour turnover. The company therefore started some organizational experiments. In order to increase satisfaction and flexibility, it abandoned scientific management and introduced principles of job rotation and group autonomy. The goal was for each shift group, including the foreman and one or two skilled maintenance workers, to develop into a relatively autonomous work group, with most workers gradually becoming qualified in most of the tasks. This development was stimulated by changes in the wage system: increased skills were rewarded by increased wages.

The experiment was a considerable success. After two years the shift groups had become very flexible. If problems in one section of the factory began to swamp the operator in charge, workers from other sections would give a hand. Training new workers in this system of organization also proved to be relatively simple. After a short introductory course, the recruit could cope with simple tasks. If problems arose he would normally get help and explanation from qualified workers nearby. The new pattern of work was very adaptable to absenteeism as well; if a worker was sick, his mates could look after his job. When faced with an excessive work load, some tasks could be postponed until the next shift.

This changed the interdependence between workers and management. Before the alterations, unskilled shift workers could leave the company if the labour market permitted, and find work elsewhere without losing wages or benefits. Though the firm could hire unskilled labour, production suffered from high turnover. The changes made the workers more dependent on the firm: they acquired skills that were of little value elsewhere, and they could not expect to get equally good wages and conditions in other companies. Meanwhile the firm became less dependent on each individual worker. With a considerably more flexible work organization, the firm was less vulnerable to absenteeism and labour turnover. Providing shift workers with tools and skills to perform first line maintenance tasks had a similar effect: the strategic importance of skilled workers was reduced.

The New Skills
Rapid technological advances since the Second World War, most notably in fields like data processing, communication and instrumentation, have stimulated the development of new skills. These skills differ from the old skills in their lack of tradition and established principles of organization. Technological development is no threat to them. On the contrary, their development and prestige are enhanced by scientific and technological innovations. As with the old skills, they seem in demand virtually everywhere in modern industry, commerce and administration.

Electronic data processing is the new skill *par excellence*. The computer industry is still dominated by young people, the large majority of whom probably have little or no prior working experience in other fields. Finding computer work interesting and handsomely paid, many people have interrupted university studies or alternative business careers and settled down as computer experts without any formal education in the field. Though quite a few still enter this way, things are changing. In many countries computer training has entered the school system at all levels, and training procedures are becoming standardized and formalized. Nevertheless, rapid technological change limits this formalization.

So far the computer industry has expanded rapidly and been very profitable. In order to get qualified people, companies have had to pay high salaries. Skilled but dissatisfied computer personnel have usually been able to find new and often better paying jobs without difficulty. With good opportunities to make individual careers, to start up their own businesses or to operate in a labour market where sellers have the upper hand, few computer personnel have attempted to organize trade unions. There are other reasons why new skilled workers tend to prefer individual strategies. There are no formal criteria specifying standards of performance. Unlike the traditional skills there is a vast difference between the excellent and the mediocre. The best people are at the forefront of technological development, and many enter the field because they find technological puzzles stimulating. All in all, the computer industry is apt to have a very individual-centred working culture.

Nevertheless, if computer specialists attempt collective action, their high degree of individual independence give the trade union a very strong bargaining position. One problem facing such unions is that many employers are willing to offer the most skilled people higher salaries than a union can negotiate. Common interests in discussing technology, work ethics, etc. will most likely give such unions the character of professional associations rather than traditional trade unions.

Unskilled Workers
This is a large and heterogeneous category. Some unskilled workers, such as dockers or car workers, have managed to organize strong trade unions. But unskilled workers generally have a weak position (Stearns, 1976). They cannot use skills to shelter their position in the labour market, and they lack the social basis which common training may provide.[1]

Unskilled workers can move, but can also be replaced, in the part of the labour market which is not sheltered by skills. This gives them a good strategic position when the economy is buoyant, but in a recession, conditions often change dramatically for the worse. There is nothing to shelter them from competition with the unemployed, secondary work force: women, immigrants, young people etc.

Belonging to a labour market with few segmental barriers, unskilled workers tend to move a lot. If dissatisfied with a job, they can move to another firm without losing special benefits. One job is as good or as bad as the next. Some of the movement of the unskilled labour force is brought about by employers. Having invested little in training unskilled workers, and being able to replace them easily, there is little to prevent an employer from firing superfluous workers unless there exist strong trade unions.

High turnover among unskilled workers makes it hard for unions to gain a foothold. In many workplaces we find a vicious circle: with a weak trade union it is difficult for the workers to improve wages and other conditions in the workplace. Rather than engage in union activity, many workers decide to quit; this undermines the development of unity in the workforce. Some workers do stay and try to organize a union; but with low membership and a high turnover, the union is faced with the unending task of recruiting new workers; and the fact that it can claim few positive results makes the task even more formidable.

Nevertheless in some workplaces unskilled workers manage to organize effectively. Such union activity is likely to reflect their independent situation. In strong contrast to, for example, worker bureaucrats, unskilled labour has little to lose from oppositional activity, having no special benefits or promotion opportunities. Thus militant behaviour may be expected.

Summary

Skills are the causes and effects of differences among workers. The preceding analysis indicates a number of relationships.

Craft workers have a broad range of options. Opportunities to exit make them independent and likely to speak out, both to supervisors and at trade union meetings. Individual independence and collective attitudes and traditions provide for strong unions with control of their segment of the labour market. This creates a strong bargaining situation for the members, making them relatively independent of the employers. This individual freedom in turn adds to union strength.

While exit is almost entirely out of the question, individual as well as collective voice are important options for *worker bureaucrats*. Generally supporting trade unions, they tend to avoid radical acts that may threaten the company and the job.

Lacking transferable skills, the *company specialists* are highly dependent of the employer. Exit is a very unattractive option, and they will tend to avoid confrontations of any kind, whether individual or collective. Union support is likely to be weak. This leaves the company specialists with very narrow options. Most probably they will try to make a career in the internal labour market.

Individual options of various kinds, be it voice or moving in the internal or external labour market, are open for the *new skilled workers*. Though often very attractive, such options are likely to prevent collective action. If they unionize, the unions will probably reflect the individualism among the new skilled workers: their bargaining position is likely to be strong, but they may lack broad social support.

Unskilled workers are prey to the labour market. Exit is an attractive or difficult option depending upon the market. Exposed to competition from the unemployed, unskilled workers generally have a weak bargaining position. The collective potential of unskilled workers largely depends on conditions not discussed in this chapter.

	Craft workers	Worker bureaucrats	Company specialists	New skills	Unskilled workers
Exit	+	−	−	+	+
Individual voice	+	+	+	+	−
Collective voice	+	+	−	−	−

Trends

The concepts introduced in this chapter allow us to specify the effects of a number of different trends on the strategic position of workers.

The Effects of Monopolization

Though the major aim of industrial mergers is to improve the strategic position of firms in the markets for capital, raw materials or products, they have effects in the labour market as well. Monopolistic tendencies imply fewer options for workers in the labour market. Turning general into company specific skills, such tendencies may weaken workers' bargaining position significantly. Relative growth of small and medium size companies has the opposite effect.

Managerial Strategies for Work Design

Friedman (1977) has made a distinction between two types of managerial strategies in work design: direct control and responsible autonomy. The first is a continuation of Taylorist principles and implies the replacement of skilled workers in strong bargaining positions and with predominantly collective orientations by weaker semiskilled and unskilled workers. This tendency is probably still quite pronounced. The strategy of responsible autonomy often entails that workers become company specialists. This strategy corresponds to the 'Scandinavian model' and the QWL approach, as illustrated by the example from Norsk Hydro.

This chapter suggests that both of these strategies weaken the strategic position of the workers, but the latter is possibly more dangerous for the workers, double-edged and less easy to penetrate. The example from Norsk Hydro shows that when unskilled and skilled workers are replaced by company specialists, job satisfaction increases among the majority of the workers. However, the change makes the workers more dependent on the company, and more oriented to individual careers than to collective action.

Integration of Science in Production
This can make it difficult to defend the social construction of skills. New technology will make old skills obsolete, and reduce the number of worker bureaucrats and craft workers.

Policy towards scientific and technological development is a controversial issue among trade unions. To the extent that economic competition is limited to one country, strong local and national unions may have sufficient strength to maintain social definitions of skill and thus defend the strategic position of the workers.

In order to grasp the dynamics of the current situation, however, an international perspective is necessary. In leading industrial nations, strong trade unions may be able to resist the adverse effects of technological change and maintain the social definition of skill as long as their industry is technologically and economically at the forefront. However, as time passes, employers in such countries may be prevented from making technological advances so that gradually they will be overtaken by competitors in countries where labour is weak. This, in turn, is likely to ruin the bargaining position of the trade unions and reduce their strength. British industry has probably developed along these lines.

The picture changes when we move to industries where the trade unions of the leading industrial countries are weak or non-existent. In this case technology may develop rapidly and without serious social constraints. Trade unions in secondary industrial nations that try to defend their social definitions of skill, are liable to reduce the competitive strength of their industry and destroy their own bargaining position. According to this argument, one of the greatest obstacles for European trade unions is the weakness of labour in the two leading industrial powers, the USA and Japan. Attempts to strengthen the unions in these two countries should therefore be strongly supported by unions elsewhere.

Conclusion
Having identified some possible trends, we have seen that they may have different and frequently contradictory influences on the development of skills. But anticipating increasing monopolization, a continuation of the present trends in work design, and finally rapid scientific and technological transformation at an international level, the following outcome is likely: the relative number of craft workers will be reduced dramatically, while the number of new-skilled workers and company specialists will increase. The size of the remaining category, the worker bureaucrats, depends on the expansion of the public sector and is hard to estimate.

The skill category which has traditionally been the backbone of

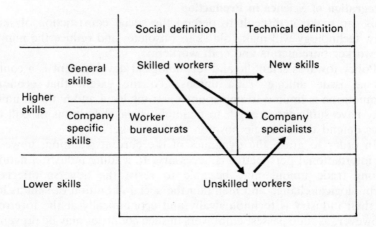

collective action and unionism is likely to diminish. Categories inclined toward individual strategies, and probably lacking collective potential, are likely to grow rapidly. This prediction which reflects a Weberian approach to class analysis, differs from Braverman's by emphasizing the increasing variety in strategic positions for different types of workers. The concepts elaborated in this chapter, by moving beyond traditional one-dimensional definitions of skill, expose developments which would otherwise remain invisible.

Note

1. The great differences between various groups of unskilled workers indicate that lack of skills is only one of several characteristics of this body of workers. Other determinants of worker power may have a significant impact as well, but this discussion is limited to the ways in which lack of skills affects the strategic position of the workers.

References

Altmann, N. and F. Bøhle. 1972. *Industrielle Arbeit und Soziale Sicherheit.* Frankfurt a.M.: Athenæum.

Braverman, H. 1974. *Labor and Monopoly Capital.* New York: Monthly Review Press.

Clegg, H. 1976. *Trade Unionism under Collective Bargaining.* Oxford: Basil Blackwell.

Doeringer, P. B. and M. J. Piore. 1971. *Internal Labor Markets and Manpower Analysis.* Lexington: Heath.

Emery, F. 1959. *Socio-technical Systems.* London: Tavistock Institute of Human Relations.

Emery, F. and E. Thorsrud. 1976. *Democracy at Work*. Leiden: Nijhoff.

Friedman, A. L. 1977. *Industry and Labour*. London: Macmillan.

Gulowsen, J. 1974. 'Bedriftsdemokrati ved Norsk Hydro?'. AI dok 4/74 Oslo: Arbeidsforskningsinstituttene.

Hirschman, A. 1970. *Exit, Voice and Loyalty*. Cambridge, Mass: Harvard University Press.

More, C. 1980. *Skill and the English Working Class*. London: Croom Helm.

Stearns, P. N. 1976. 'The Unskilled and Industrialization'. *Archiv für Sozialgeschichte*. Vol. 16, 249–82.

10

New Technology, Training and Union Strategies

Helen Rainbird

As a result of contemporary developments in the labour market it is not always easy to separate union responses to the acquisition of new technology skills from responses to employer demands for greater labour flexibility. In principle they are distinct, but in practice they are interconnected, and flexible working practices have been presented as though they have a technological imperative. As a result, it is often difficult to distinguish managerial policy towards training for new technology (or rather, its absence) and policy towards training for increased flexibility (or its absence). Employers have a degree of choice in the way in which production is organized and jobs designed. In Britain, it has been argued, management pursues one of two strategies: on the one hand, tasks on mass production lines are compartmentalized so that the worker acts as an extension of the machinery, exercising no control over production decisions (cf. Braverman, 1974). On the other hand, in small batch production, workers have retained more autonomy in design decisions because of the greater flexibility of product lines (Friedman, 1977). An outgrowth of the latter concept has been the association of craft skills and labour flexibility with increased worker control over the labour process. Whilst this may have been true in the firms studied by Friedman in the 1970s, the demands placed on workers of all grades to broaden their range of work tasks in the recession of the 1980s may result neither in an upgrading of skills nor an increase in work autonomy.

Trade unions recognize that companies invest in new technology to remain competitive. Their policy is therefore not to resist technological change, but to attempt to control its introduction in seeking agreements on new technology, disclosure of information and job security. In doing this, a major concern has been to obtain greater involvement in labour force planning, and hence a direct interest in matters relating to training and retraining. The acquisition of new skills and the defence of jobs against deskilling relate directly to conventional union concerns with wages and conditions.

Unions recognize that there are positive and negative effects of new technology on jobs, and invariably stress the need to seek the positive benefits in negotiations. However, there are divergent processes at work. Some workers are enjoying improved conditions of employment, greater job security and, by becoming more flexible and acquiring new skills, arguably becoming more skilled. Others are still employed but are finding their work deskilled and degraded. But undeniably, new technology introduced in the context of industrial restructuring in the recession has been a contributory factor towards massive job losses. This, combined with the casualization of large sectors of the working population through firms' use of part-time, short-term contract and subcontracted labour has resulted in both a degradation of work and a deskilling process through loss of skills at the societal level.

The Concept of Skill

Braverman (1974) argues that the development of the productive forces under capitalism requires constant technical innovation whereby management seeks to extend managerial control over the labour process by separating conception from execution of work. *Labor and Monopoly Capital* has generated debate on the extent to which scientific management techniques result in the deskilling of labour and the degradation of work (see chapter 8). Here, it is useful to refer to Cockburn's study of the printing industry in which she distinguishes between different types of skill. She argues that loss of skill is distinct from degradation of work and also from loss of control. Skill itself is a multi-faceted concept which refers firstly to accumulated experience; secondly, to the skill demanded by a particular job; but thirdly has a political dimension in so far as a group of workers or a union defends its skills against the challenge of the employer or another group of workers (1983: 113). The combination of these three elements can be clearly seen in her examination of the skilled work of the printers.

Discussion of the nature of craft skills must be set alongside the assertion by feminists and general unions that all workers have skills, but they have never been recognized through the wages structure as have been those of skilled craft workers. Phillips and Taylor (1980) argue that the classification of jobs as skilled or semiskilled has more to do with the sex of the worker than the training or ability required to perform them. Employers do indeed differentiate their workforces by gender and race and this has been documented in the literature on labour market segmentation (e.g. Wilkinson, 1981). Despite this observation, there are real differences in the level of skills demanded by particular jobs which have more to do with the way the job is designed and how the worker is prepared to perform it than with the innate abilities of the worker. There is a fundamental contradiction between the productive potential of human ability and its restriction to specific job tasks in capitalist labour processes which are differentiated by their real and supposed skill content.

When the complexity of the concept is 'unpacked', it is possible to see that there is no contradiction between management's objective of increasing the division of labour and compartmentalizing tasks in the labour process which result in a degradation of work, and the fact that, regardless of whether skills are explicitly rewarded and recognized through job hierarchies and the wages system, workers have accumulated experience and knowledge of work processes which they apply in the execution of their jobs. Whilst managerial strategies of control degrade and deskill work, workers struggle to re-exert control. Since under capitalist relations of production many existing skills are not formally recognized or allowed to develop, there is no reason why new technology labour processes will not be organized around jobs which require formally recognized skills and jobs which, like those required for earlier generations of compartmentalized and degraded work, do not acknowledge the skills and abilities of their occupants.

In the context of the debate on labour flexibility, skills are most likely to receive recognition in the wages structure if unions effectively claim and defend them. The problem is that management may impose new skills on workers and resist acknowledging their existence if training is very short or informal and takes place outside the ambit of union control and initiatives. Lee argues that

> the interesting question . . . is not simply the issue of 'objective' changes in the skill content of jobs brought about by the industrial locus of employment; it also involves the issue of whether worker organization is able to ensure that the new expertise is actually recognized (1982: 154).

Therefore the unions' ability to claim new skills and to obtain formal recognition for them will be fundamental to the redefinition of skilled work.

In a period in which new technology is altering the nature of work and flexible working practices are breaking down established job demarcations, the claiming of new technology skills takes on a significance for unions as organizations. Cockburn distinguishes the individual worker's sense of satisfaction in exercising skill from the meaning it has for organizations representing collective interests. She argues that whilst there may be collective struggle *over* skill, it may not necessarily be *about* skill, but rather its market value (1983: 121). Nevertheless, in the present context, new technology is undermining the distinctive identity of union organizations by breaking down distinctions between white-collar and blue-collar work, between unskilled and skilled work and between trades. Therefore the claiming of new skills and occupations will be decisive to future developments in the union movement.

Managerial Control and the Degradation of Work

New technology based on microelectronics is characterized by its cheapness and its wide applicability compared to earlier generations of technology (TGWU, 1979). Information technology, in particular, has an enormous capacity to integrate systems and to exercise centralized control over the pace of work and in monitoring performance. This centralization of control cuts out the need for human labour and removes decision-making in many of the jobs that remain. Not only are jobs changed with new technology but the centralization of information allows peaks and troughs in labour demand to be monitored precisely and so allow part-time and temporary labour to be deployed effectively and efficiently. If this information system is linked into robotic controls in warehouses and laser beam checkouts in retail outlets, then the potential for both degrading and deskilling becomes apparent. In interviews, officials in eight major unions with membership in chemicals, engineering and food processing reported the following developments.[1] On part-time working, they commented, 'women are being turned on and off like a bloody tap' and that increasing usage of part-timers revolved around 'the economics of the tea-break'. They were emphatic that jobs are being deskilled:

> You don't need skills to pick up a can of beans and wave it over a laser beam (USDAW).

and

> With robot pickers in warehouses there will still be jobs for USDAW members, but they will be unskilled and low paid, with the physical and often the thinking part taken out. We are concerned with what job will be left after new technology is introduced (USDAW).

Not only does new technology destroy and degrade unskilled and semiskilled work, skills performed by craftsmen are not invulnerable to technological redundancy. Machines introduced in press shops eliminate the need for tool changes, a highly skilled job. A GMB official went so far as to argue that management strategy is to create stations which require minimal training and, by eradicating work which requires training, to eradicate skill and higher pay. Another strategy to reduce the need to employ skilled people in engineering is by transferring skilled cutting work from the factory to the steel stockholders. In this way, the problem of managerial control, 'man management', is removed from the labour process altogether as the discipline of market competition is imposed on the suppliers. Computer numerically-controlled (CNC) machine tools, whilst enhancing the individual toolmaker's job by taking on part of the design function, decrease the overall demand for skilled toolmakers. 'For the unions, new technology doesn't provide any members. New machine tools don't pay membership dues, attend union meetings, get sick or take holidays' (AEU).

The major means by which managerial control is extended is through new working practices in conjunction with, but independently of, new technology. Occupational flexibility is imposed on all grades of workers from unskilled and semiskilled through to craft workers. Though it can concern only the peripheral areas of skills, there have also been moves towards complete interchangeability of tasks. Employers appear to be seeking 'a general worker philosophy whereby all workers are general and flexible'. GMB officials in the Midlands felt that changes had occurred in jobs and members had taken on new skills without the unions responding properly. This represented a development of flexible attitudes towards changing job definitions and boundaries, which will serve as a preparation for a major assault on craft skills and demarcations at a later stage.

In a period of high unemployment, when workers themselves are under the threat of redundancy, pressures to adopt flexible working practices can be intense:

> Companies rely on the fear of the workers that if they don't adapt and exhibit flexible attitudes and abilities of their own innovation then they are ripe for selection for redundancy. Incredible pressure is being placed on those of 45 and 50 who dread what's happening and the indications suggest that older people are accepting menial tasks to avoid being trapped by constant changes of semiskilled jobs. Janitorial jobs are at a premium.. . . At the same time the monotony and boredom of screwdriver assembly work is tremendous because the skills are so watered down (GMB).

This raises the question of how workers become more flexible and it appears that in many instances the process of learning the skills required for a new job occurs through informal training by one operator training another. This suggests that new skills are being acquired without the legitimation of a formal training period. Therefore, although the scope of a job is being extended, perhaps even taking on some aspects of semiskilled and skilled work, this is not always rewarded in the wages structure. New skills are being imposed but not recognized.

Failure to recognize new skills not only occurs where training is organized on an informal basis, but also where training is narrow and competence is acquired largely through experience. This has occurred in office work with the acquisition of new technology skills in word processing. Typists are sent on one- and two-day courses run by the manufacturer, which are specific to the machine. APEX advises secretaries undergoing manufacturers' courses in word processing to realize that full competence can only be acquired after 3 to 6 months experience in addition to the introductory course. The union argues that there is a need for formal qualifications in the principles of word processing (i.e. transferable knowledge and skills) and recognizes the weakness of many agreements which were made on new technology in the 1970s which concentrated on job security and pay. In failing to raise issues relating to retraining and job design, many word processor operators have encountered difficulties in obtaining regrading since they cannot demonstrate that their jobs require more skill than previously (APEX, 1980).

An example from the food processing industry of reorganization and investment in new technology indicates how management may fail to recognize the value of skills of a supposedly unskilled workforce. This company decided to close down several processing and packaging plants on one site whilst introducing new bottling, warehousing and production systems. It called for voluntary redundancies whilst offering retention payments for those they wanted to re-employ when the new buildings were opened. However, in their haste to let people go, they lost skills which were essential to production. These skills, though not formally recognized, had been acquired through experience and were based on knowledge of the system, the product and an ability to identify faults in the functioning of production lines. Although the jobs were considered to be semiskilled and requiring manual dexterity rather than diagnostic skills ('the old argument about women having round bottoms and being good at doing a job which requires sitting all day'), the company had to bring back these 'unskilled' women workers who had been made redundant, many of whom had subsequently found alternative employment, to train the new employees. These skills are tacit because, like imposed skills, they are not acknowledged.[2] The example demonstrates that despite management attempts to extend control through the division of labour and the categorization of certain types of work as requiring little or no skill, workers reskill the intellectual and manual content of their jobs autonomously. As Marx commented:

The worker's continual repetition of the same narrowly defined act and the concentration of his attention on it teach him by experience how to attain the desired effect with the minimum of exertion. But since there are always several generations of workers living at one time, and working together at the manufacture of a given article, the technical skill, the tricks of the trade thus acquired, become established, and are accumulated and handed down (1976: 458).

It is precisely this 'untapped energy and skill of . . . workers who have learnt from their own experience' (TGWU, 1979: 5) that quality circles attempt to exploit and which could form a means both of increasing productivity and of dignifying work if workers exercised real control over the labour process.

Skilled Workers and New Technology

If the skills of unskilled and semiskilled workers have often been unacknowledged, then this is not the case for craft workers. Entry to distinctive trades has historically been through time-serving and 'sitting next to Nellie' though since the 1964 Industrial Training Act this has commonly been by means of formal, off-the-job training with courses set up through the Industrial Training Boards.[3] It is amongst craft workers that the political and organizational component of skill has been most apparent, and which has given rise to employers' complaints of inflexibilities. Of course, the unions argue that employers themselves have created inflexibilities through breaking jobs down into their component parts and that the unions simply defend their members' jobs. This becomes a problem for the employer when it appears desirable to change job design again. Whilst employers' demands for labour flexibility are not new, its potential to increase productivity with new technology is enormous.

So what are the implications of increasing flexibility for the skills of craft workers? Firstly, a distinction has to be made between flexibility in peripheral areas of a craft and multiskilling, otherwise known as 'skill exchange' or 'third streaming', whereby the jobs of distinctive crafts are rolled up together so that, for example, an electrician or a fitter could do both jobs. That is to say, that craft skills become interchangeable. The former is widespread and, it could be argued, the general unions and some of the craft unions, who would argue that their members have always been 'all-rounders', would see it as a means of enhancing skills. Whether this or multiskilling enhances skills or reduces them is open to debate. Compared to concessions made on flexibility in peripheral areas of craft skills, the impact of multiskilling is more far-reaching and there is greater dissonance between what it might mean for the skill content of individual jobs and its consequences for organizational identity. This accounts for the unions' concern for the social cost of the massive increases in productivity and concomitant redundancies that multiskilling will bring about. More importantly, multiskilling opens up a hornet's nest of potential disputes by undermining the distinctive occupational basis of each union. In chemicals,

> technological changes didn't bother us in the past but the new plant is more sophisticated and computer-controlled and has resulted not so much in conflict but in abrasions with other unions which didn't exist before. This is intensified by unemployment and the recession. We are finding inter-union rivalry quite intense in many places and it is severely straining the sense of brotherhood in the movement (AEU).

Whilst not all unions see the spectre of multiskilling on the immediate horizon there are already formal agreements on multiskilling in a number of major companies, though many of these are on greenfield sites (Incomes Data Services, 1986). Moreover, the most astute observers of employer strategy in chemicals consider that concessions on flexibility in the peripheral areas of craft skills have paved the way for major initiatives on multiskilling. So potential conflicts between craft unions are opened up, as between craft unions and general unions, representing process workers.

The unions' views on whether multiskilling results in upgrading or degrading of skills varies but, regardless of its effect on members' jobs, they have an interest in justifying their positions independently of the objective situation in an effort to retain membership and spheres of influence. Nevertheless, the blurring of demarcation lines, and the homogenization of labour raises important issues for the union movement. Firstly, whilst competition for members can increase conflicts between unions, it also makes unity a pressing objective. Therefore, whilst new technology has the potential to open up disputes as traditional boundaries between jobs become blurred, unions can respond by moves towards amalgamation. The latter will be brought about not just by employer initiatives on flexible working practices as outlined above, but also by the technical aspects of new technology, which both proletarianize white-collar work and undermine existing concepts of skilled and unskilled work. Secondly, new technology makes the human element in production increasingly dispensable.

> Old activities are disappearing faster than new ones are generated. We fear that if this is not regulated then the future will be bleak for the job expectations of many people. Job reductions in chemicals have coincided with the highest level of production

ever. This indicates the extent of the problem in accommodating new training. The unions have been remarkably accommodating because their members are being dispossessed, whereas business accepts every advance in technology because it has to compete for markets (AEU).

When unions talk of the 'social cost' of new technology they are normally referring to the cost to their members made redundant, arguing the case for adequate compensation. However, the societal cost of casual work and unemployment for large sectors of the population is not confronted by the sectional interests of the unions.

Claiming New Technology Skills

Apart from defending traditional skills, unions actively claim new technology skills. Given a declining population of union members and a loss of distinctive craft or occupational identities, the claiming of new skills becomes part of a strategy for organizational survival. One of the unions which has made advances in this field is the electricians' union, the EETPU. In a radical departure from previous union practice in Britain, which has been to see the provision of industrial training as the responsibility of the employer, the EETPU sees the provision of training in electronics skills as central to its image as the union for members with new technology skills and as a proponent of the 'new realism' in industrial relations. The jobs of electricians have indeed been affected by the introduction of electronics, especially in maintenance work and in the commissioning of new plant.

In 1980 the EETPU opened its own training school of Cudham Hall, Esher, where short courses in electronics, electronic logic systems, programming and microprocessor interfacing are run. The college also has mobile instructor units so that courses can be provided in plants, and in 1985 the union extended its training facilities to regional offices. There are advantages to members if they can increase their bargaining power by upgrading their skills and hence their employability by attending. Whilst other unions have continued to take the view that training is the responsibility of the employer, the AEU and the craft sector of TASS are now also involved in organizing electronics courses for their members.

As yet the EETPU is the only union to have an internal training representative structure on a par with the health and safety representative structure. The purpose of this structure is not just to raise awareness amongst members for the need to acquire new technology skills:

Training representatives would encourage the better use of

manpower availability, avoid deskilling, promote youth training, adult training and retraining.

Training representatives would promote the policies of the EETPU in new technology and industrial training, further expand our powers of influence to existing and potential members and provide a marketing force for our own technical training facilities (EETPU *Training Bulletin*, March 1983, p. 8).

For the EETPU, training in new technology skills stakes a claim to the skilled work of the future and constitutes a means of expanding influence amongst other craft workers. The strategy of TASS, another union with claims to high technology skills based on its membership in drawing offices and amongst technicians, is to expand through amalgamation with craft unions: the rationale being new technology's capacity to break down the barriers between craft and white-collar work. This has led to a massive expansion in membership, and although amalgamation with the AEU was eventually called off in 1985, TASS has added four craft unions to its craft sector since 1981. Part of the success of this strategy has been in the link between high technology skills and increasing occupational status:

It is this belief that merger with TASS will place a small union on the right side in the new technology battles plus increase its members' chances of acquiring staff status that has been decisive (*Financial Times*, 29 May, 1985).

Clearly, secure jobs in the enterprise of the future will be those which have the attributes of skill to which unions such as TASS and EETPU have staked out a claim. However, organizational claims to represent a new technology-skilled membership may be at variance with members' feelings about what is happening to the skill content of their jobs, which concerns how it relates to their accumulated knowledge and the demands of the job itself. These union members will exercise their skills in the context of specific labour processes, subject to managerial control. The extent to which the acquisition of new technology skills results in deskilling and the degradation of work can only be tested by empirical study of specific labour processes.

Finally, new technology does not just reduce employment and change the nature of work in traditional industries, but also creates new industries with recruitment potential for the unions. Although these are not labour-intensive and are unlikely to replace the numbers lost elsewhere, strategically it is important for the unions to make inroads here, especially for those with new technology claims.

Conclusion

New technology is not neutral and has the potential vastly to expand managerial control over labour processes. There are high skill and low skill options in managerial choices over its introduction and clearly the extent to which the high skill, high productivity option is taken will largely be determined by the extent to which unions take initiatives in claiming new skills and obtaining recognition for them. Formal training programmes and plant-based involvement in job design are major considerations in this. Involvement in manpower planning at plant level may result in greater identification with the interests of the firm than has previously existed and employers' pressures for single site and single union agreements would indicate that this is a conscious strategy on their part. However, workers who benefit from secure employment and high technology skills do so in a general context in which there is an increasing polarization in the division of labour. Jobs are being casualized and new skills imposed without receiving recognition. In this chapter, union perceptions of the impact of new technology on members' jobs have been examined and strategies towards defending and claiming skills. These strategies have implications not just for the individual members but for unions as organizations with membership territories to defend. Whilst a declining base of fully employed members, combined with the breaking down of existing divisions within the labour force may open up sectional conflict over spheres of influence, it also poses its antithesis, co-operation and amalgamation. There can be little doubt that the introduction of new technology and attendant changes in occupational structure will see new alignments in the trade union movement in the coming years.

Notes

The author wishes to acknowledge the support of the ESRC for this research.
1. Interviews were conducted with national officers and research staff at national level and officers with regional responsibilities at local level in the Association of Professional, Executive, Clerical and Computer Staff (APEX), the Association of Scientific, Technical and Managerial Staffs (ASTMS), the Amalgamated Engineering Union (AEU), the Electrical, Electronic, Telecommunication and Plumbing Union (EETPU), the General, Municipal, Boilermakers' and Allied Trades' Union (GMB), the Technical, Administrative and Supervisory Staffs (TASS), the Transport and General Workers Union (TGWU) and the Union of Shop, Distributive and Allied Workers (USDAW).
2. I am grateful to Peter Armstrong of the Industrial Relations Research Unit for suggesting these terms to me.

3. The Industrial Training Boards (ITBs) were tripartite bodies with statutory powers to raise a training levy on all firms defined as coming within their scope. They set standards for training and were also involved in course design. As a result of the 1981 Review of the Employment and Training Act, 17 of the 23 ITBs were abolished, though two of the largest boards – engineering and construction – remain (see Rainbird and Grant, 1985, for details).

References

APEX. 1980. *Automation and the Office Worker*. Report of Office Technology Working Party. London: APEX.

Braverman, H. 1974. *Labor and Monopoly Capital: The Degradation of Work in the Twentieth Century*. New York and London: Monthly Review Press.

Cockburn, C. 1983. *Brothers: Male Dominance and Technological Change*. London: Pluto.

EETPU. 1983. *Training Bulletin*. March.

Friedman, A. 1977. *Industry and Labour*. London: Macmillan.

Incomes Data Services. 1986. *Flexibility at Work*. Study No. 360, April.

Lee, D. 1982. Beyond Deskilling: Skill, Craft and Class. *The Degradation of Work? Skills, Deskilling and the Labour Process*. Ed. S. Wood. London: Hutchinson.

Marx, K. 1976. *Capital*. Volume 1. Harmondsworth: Penguin.

Phillips, A. and B. Taylor. 1980. 'Sex and Skill: Notes towards a Feminist Economics.' *Feminist Review*. No. 6.

Rainbird, H. and W. Grant. 1985. 'Employers' Associations and Training Policy. Industrial Training Arrangements in Four Industries: Food Processing, Chemicals, Machine Tools and Construction'. Institute for Employment Research Report, University of Warwick.

TGWU. 1979. *Microelectronics. New Technology, Old Problems. New Opportunities*. London: TGWU.

Wilkinson, F. (ed.) 1981. *The Dynamics of Labour Market Segmentation*. London: Academic Press.

Part IV

Trade Union Strategies

11

New Technological Paradigms, Long Waves and Trade Unions

Otto Jacobi

The current phase of dramatic technological innovation coincides with radical political changes: seen by some as the 'end of Keynesianism' or the 'collapse of social-democratic consensus'. The argument of this chapter is that technical and political developments are causally related. However, the connections are complex; and the very speed of technological change makes the identification and extrapolation of its social consequences increasingly difficult.

Rather than attempting to predict the future, the best that I can do is to suggest alternative scenarios, each in principle plausible. This chapter aims to present – as a 'counter-scenario' to the pessimistic expectations predominant in industrial sociology and industrial relations – some developments which point to a more favourable future. This is not to dismiss those scenarios which emphasize the potential for destruction, dominance and manipulation accompanying modern technology and which take a sceptical view of the development of growth, employment and the world economy. Such models are of great heuristic value as they play an important part in recognizing negative developments and correcting them. Therefore, we would be well advised to take seriously Gunter Anders' attack (1984) on 'the happy end-politicians and happy end-publicists who are not ashamed of dealing in optimism'. In presenting here a more cheerful, optimistic scenario, it is recognized that what is submitted is no more, but also no less than a future perspective founded on substantial evidence.

The basis of such a scenario is the judgement that new generations of technology are conducive to growth and wealth, ecologically innocuous and economical of resources. It assumes that on the basis of 'new radical innovations', new products and new production methods and also extended and worldwide interdependent markets will come into existence, leading to a long-term economic upturn and contributing to increased individual and collective welfare and to a limitation of environmental pollution.

The political concomitants of a technologically-based economic upturn can be seen in totally opposing terms. For conservatives – (e.g. Kahn, 1983) – state intervention and co-operation with trade unions have intensified political control and social regulation to such an extent as to become obstacles to innovation leading to institutional sclerosis. The solution is to extend entrepreneurial freedoms and release market forces. In contrast there is a view – not only on the left – that transition to a new level of capitalist development requires political regulation: technology and economic organization must be geared to social needs, and therefore increased political intervention can be expected. These opposing scenarios may be starkly identified as 'creative destruction' and 'creative altruism' (Matzner, 1982; Meissner and Zinn, 1984).

Both variants of the optimistic scenario agree that capitalist democracies currently face dilemmas of transition: a crisis of development but not of the system.

Both also assume that the consequences of technological innovation do not emerge spontaneously but are socially conditioned. The key difference is in the nature of this social mediation.

The following discussion seeks to present evidence for the propositions:

1. that long-term economic expansion is more likely than sustained recession;
2. that an extension and enlargement of state economic intervention is more likely than a reduction;
3. that trade unions are not doomed to a merely passive, reactive adaption process; new fields of activity are open to them which allow innovation through change and will secure them a future presence on the economic and political bargaining markets;
4. that there is emerging in all capitalist democracies a polarization between a model of labour exclusion and a participatory model.

The Coming Upswing

Predictions about long-term economic development have increasingly been corrected upwards in the last few years. The International Bank of Reconstruction and Development forecast in April 1984 that average annual growth between 1985 to 1995 will reach 3.3 per cent in the industrialized countries, 5.1 per cent in the developing countries and

5.8 per cent in the newly industrializing countries. Forecasts by several economic research institutes for the Federal Republic of Germany speak of 2.5 per cent or 3.5 per cent growth. For the USA, Kahn (1983) predicts economic growth of 3 per cent or more, which would mean a doubling of the gross national product between 1982 and 2002. These forecasts assume that the growth potential of new technology can be realized and dislocation of international goods and credit markets can be avoided. In view of the high gains in productivity, there will be a marked improvement in the labour market only if the number of persons seeking employment decreases or working hours are reduced drastically.

These economic projections reveal surprising agreement among social scientists with very different political and theoretical orientations.

Two left-wing German economists, Meissner and Zinn (1984), argue that 'The New Prosperity' – as their book is entitled – can only be realized if the capacity for innovation, qualitative growth and reduced working time can be harmonized. They advocate a combination of Schumpeter's innovation dynamic and Keynesian demand policy, and consider national consensus and international co-operation both necessary and possible. Especially for Central Europe they see an opportunity for 'intelligent interventionism' which includes participatory co-operation between the state, employees and business, and which permits a European middle course between Japan's industrial feudalism and American market conservatism. Their view partially coincides with that of Galtung (1985), who expects 'golden years' for the end of this decade and anticipates beneficial change in Western countries if they accentuate their 'softer aspects'. The French social scientist Piatier (1983) expects a new phase of prolonged growth to begin in about 1990. The American Marxist-orientated economist Wallerstein (1973) predicts for the 1990s a 'renewed marked upturn of the world economy'. Kahn, a conservative American social scientist, holds that the development of a 'super-industrial' world economy will be possible before the end of the century.

The central question for any economic prognosis is, does the emergent new technology constitute a new industrial revolution? And if so, will the consequences be similar to those of previous technological revolutions?

One distinctive feature is the fact that technological development first improved human physical power, while it now involves a gigantic extension of mental capacities. 'The present industrial revolution has totally different dimensions, since the computer revolution offers humanity an augmentation of its intellectual forces' (Piatier, 1983: 230). 'The first industrial revolution was primarily based on the substitution of muscular power first through the steam engine and later through electrically driven machines; the second industrial revolution involves the equipment of machines and production systems with information and computerized intelligence' (King, 1982: 18). Blackburn *et al.* (1985) distinguish between the stages of primary, secondary, and tertiary mechanization: 'namely the mechanization of transformation, of transfer and of control' (29).

Another classification differentiates between a first industrial revolution starting in 1780 based on the steam engine and textile industry, a second industrial revolution since 1850 with electricity as the basic technology and a third microelectronic revolution. This links to theories of long waves, according to which previous industrial-capitalist development can be divided into four phases, each comprising a long upturn and downturn cycle. This theory of long-term or Kondratieff cycles is an attempt to represent the interdependence of technology and economy in industrial-capitalist societies. Apparently the long-term pattern of development is cyclical in nature (Galtung, 1985) and – at least according to Wallerstein (1983) – the Kondratieff cycle with its expansion–stagnation pattern over 50–60 years is so far the most convincing attempt to integrate the economic and social development process within the context of 'new radical innovations' in a plausible framework of interpretation. Coombs (1984) confirms that a coincidence between the three types of mechanization and the long waves can be proved relatively well for the past (see table 11.1).

Theories of long waves can be supplemented and enriched with the concept of the product-cycle (Borner, 1981) and that of development stages of capitalist societies (Berger, 1983). New technological innovations undergo four periods of life: 'infancy with weak growth, adolescence with rapid growth, adulthood with lower growth and maturity with zero growth' (Piatier, 1983: 227). It can be proved, however, that the prosperity stages of long waves, though a good breeding ground for the full utilization of innovation and for enhancing the chances of spreading existing 'technological trajectories', are not favourable to the establishment of new technological paradigms. However, when the stage of maturity has been reached and there are indications of overcapacity and a saturation of the market and when competition has shifted from quality to price, there is increased pressure to take up 'new radical innovations', to make them marketable, and to establish a new 'technology system'. The trough in the Kondratieff cycle, however, makes future markets uncertain, so that investment in promising technology becomes exceedingly risky. 'Perhaps this is a fundamental contradiction of capitalist technological development: the strongest pressure towards a risky shift to new technology arises just when the risks of future market developments are extraordinarily high' (Kleinknecht, 1984: 66).

This problem points to the fact that new technological paradigms are indeed a necessary but by no means sufficient condition for checking

TABLE 11.1
Mechanization and Kondratieff Cycles

Long waves	Upturn	Downturn	Technical Innovation
1. Industrial Kondratieff	1778–1813	1814–42	steam power, coal, textiles
2. Bourgeois Kondratieff	1843–69	1870–90/95	steam navigation, railways, steel
3. Neomercantilist Kondratieff	1890/95–1914/20	1920–35/40	electrotechnical and chemical innovation, petrol and diesel engine
4. Interventionist Kondratieff	1940/45–65	1965/70–85	motor vehicles electrotechnical and chemical industries
5. New cycle	1985/90		microelectronics, genetic engineering, energy, materials technologies

Source: Kleinknecht, 1984; van Duijn, in Freeman 1983.

a recessionary spiral and initiating a new period of prolonged growth. 'Virtuous circles are often as difficult to disentangle as vicious circles' (Dosi, 1983: 95). The complexity of technological and social innovations becomes obvious when we study the post-war development of Western industrial countries.

Three great social innovations, all linked to the notion of Keynesianism – state intervention in the economy, expansion of the welfare system, and trade union involvement in economic partnership – were crucial determinants which allowed the technology of the 1930s and 1940s to engender a long growth period. As a new stage of bourgeois industrial societies, a capitalist democracy with a 'mixed economy' and 'mixed polity' (Offe, 1983) developed and received its legitimation from the social-democratic Keynesian model of consensus by participation and integration of the working class. It was a model which borrowed considerably from Catholic social doctrine and was not infrequently supported by conservative popular parties and governments (Lutz, 1984: 193). The current 'transition phase', the beginnings of which must be dated back at the latest to the first world-wide recession of the post-war era in 1974, is characterized by a double structural rupture.

On the one hand it is no longer possible, on the basis of old technologies which are fully developed and can be imitated, to go on with the previous pattern of expansion. 'The maturity and decline of the major industries of the last industrial revolution; the very weak growth of new activities of the industrial revolution now beginning and which is only in its embryonic phase' (Piatier, 1983: 229) form the technological problem of the transition phase. On the other hand, the social techniques developed in the post-war era have lost their impact because the current problem is not the exploitation of existing 'technological trajectories' but the creativity and innovation which must accompany the establishment of new technological paradigms. The need for social innovations constitutes the politico-social as well as the institutional problem of the transition phase.

Future developments are indeed uncertain, but this does not mean that we are stumbling blindly into a completely unknown world. If one is aware of the continuing force of the present order and of the inertia of established structures one knows that the old society co-exists with the new for a considerable time. The synthesis of old and new entails that the benefits of new technology can only be exploited if the post-war model of intervention and social partnership is reconstituted in a democratically legitimized form.

New Markets, Intervention and Participation

Cyclical patterns involve a combination of the familiar and the new. A new long growth phase will be based upon new technological paradigms, but will also depend on continued economic intervention and social partnership. This in turn will depend on the adaptability of the protagonists and their institutions and policies.

With microelectronics gradually moving beyond its infancy, and biotechnology in an embryonic stage, two general-purpose basic technologies are available. In the long run, it can be expected that microelectronic and biogenetic technologies will combine as biotronics. When energy and material technology is added, the emergence of a 'new technology system' is foreseeable. What can be expected are high gains in productivity in production, administration, and services, as well as the resolution of many contemporary problems such as power supply, environmental pollution, soil erosion. Jänicke (1985) predicts that the following will be the main growth areas:

1. organization and communication techniques ranging from information and communications technology to new transport systems;
2. automation technology with numerically controlled machine tools, robots and multipurpose automatic machines, new inspection and

quality control techniques and the computer-based integration of the entire production process;
3. processing technology with primary emphasis on environmental protection, efficient utilization of energy resources, extraction of raw materials and recycling techniques which comprise a wide range of application from upgrading coal to solar technology and biotechnology;
4. new materials (glass fibre, glass ceramics, alloys, fibre fortified plastics, etc.).

Computerization and telematization as well as a more ecological orientation of material and process technology may well be the foundation for a next long wave, just as mass motoring, electrification, railway construction, steam navigation and early mechanization were for the previous ones (Huber, 1985).

The potential already exists for new and expanding product markets. This is not merely a question of satisfying replacement needs on a qualitatively higher level, but more significantly of satisfying hitherto neglected, suppressed or unknown needs. Crucially important are those markets which satisfy needs for collective goods. Moreover, if one takes into account that the world-wide transfer of new technology will further increase internationalization, there is every reason to assume that markets will extend geographically. The development of peripheral markets is in the interest of developing and newly industrializing countries, which must create domestic markets for their further socio-economic development, as well as in the interest of industrial nations which depend on expanding markets for their new technology. A new phase of 'external conquest' of hitherto backward regions by capitalist industrialization and market relations can be expected. Kahn's super-industrial world economy, which is not in the least an overdrawn utopia, will imply that: (a) further countries will join the group of those with procedural and structural power; and (b) that institutions with extended powers for managing and regulating world economic problems will be established. Opposition to industrialization of the Third World on the assumption that the wasteful exploitation of the natural resources of the earth violates the chances of survival for all, is not only unrealistic but also arrogant, because this would imply the permanent exclusion of these countries from the potential wealth flowing from technology. Moreover, they are unjustified if the ecologically beneficial and resource-saving potential of new technologies is utilized.

The transformation processes which new technology has brought about in the production, social and political system of each individual country, as well as in international relations, are so extensive that there will be winners and losers (among nations, industries and social groups) and changes of status will take place with destructive potential. Thus securing a renewed long wave and the regulation or compensation of disadvantages is by no means only an economic but primarily a political question. This necessitates co-operation and 'creative altruism' on a national and international level in order to neutralize this economic and military destructive potential to the advantage of all. Whether military markets are established or exclusion strategies are pursued which favour national and social imbalances, or whether ecological markets are developed and the opportunities for international co-ordination and regulation are seized in order to achieve qualitative growth and social balance, is a matter of political alternatives which open up options beyond the internal dynamic of technology. The technical conditions for a long-wave recovery already exist to a large extent, but they are not sufficient, since what matters is also political and institutional change (Coombs, 1984).

Such political alternatives are controversial. Conservatives deny the necessity and desirability of continued state intervention and social partnership. The much-discussed thesis of institutional sclerosis (Olson, 1982) is acceptable in its general version, which says that traditional structures and institutions are increasingly incompatible with new technology. In the history of technology, economy and politics in capitalist societies, industrial revolutions or long waves were always associated with political and institutional innovation. Such moderniz-ation was always accompanied by strains and conflicts, involving the replacement of old by new politico-social forces or institutions and/or their adaptation to changed conditions. Examples of such processes are the emergence and evolution of social democratic parties and trade unions on the one hand, and the constant reorganization and rearrangement of bourgeois parties and the state on the other – a development which has led in the post-war period to the establishment of participatory-consensual capitalist democracies. Changes with simi-larly far-reaching consequences will also be found at the end of a long transition period 'which can last 30 to 50 years before it leads to a completely different type of world society' (King, 1982: 31). In the meantime – approximately till the end of this century, due to the time lag of adaptation of politico-social and institutional structures to the new technology – the political culture will be characterized by group intervention and co-operation. It is assumed then that a substantial, historical break with the 'mixed economy' and 'mixed polity' is not on the agenda.

This means that the conservative response to institutional sclerosis cannot be said to have a long-term future. According to this standpoint, too much intervention and regulation on the part of the state and the large interest organizations have become obstacles to innovation so that de-collectivization has become a prime necessity. The strengthening of the state's authority through general ground-rules and a populist bypassing of functional interest representation is intended to exclude the trade unions from the political market; while the revival of market forces is aimed at weakening the trade union position at plant level and on the labour market, and at stimulating innovation. Conservatives claim that the new technology has an inherent potential for flexibility and elasticity which will open up new degrees of freedom if – by means of a substantial reduction of the existing network of intervention and regulation – a new 'social-cultural innovation stage' can be established (Landesregierung Baden-Württemberg, 1983; Späth, 1985).

In contrast to this position there are already today many indications that the state will play the decisive role in the development and diffusion of new basic technology and in the societal changes which accompany this process. The new 'social-cultural stage of innovation' will not only be interventionist, but will also continue to undermine the classic entrepreneurial role of the 'heroic innovator'. Pump-priming investments by the state especially in the field of science and research, its supportive measures such as infrastructure modernization, or the removal of political obstacles and provision of financial assistance to transform technological innovations into marketable goods, turn the state into an indispensable and also innovation-promoting partner. Economic intervention will also take on a new quality. The much-discussed information society depends on the state providing huge investments to lay the technological foundation for a completely integrated, universal telecommunications network. As far as collective goods are concerned, their markets are created only by public demand. By guaranteeing companies a ready market for their products, the state also pursues a policy of promoting and selecting technologies. Since a long-wave recovery has a worldwide dimension, the maintenance and improvement of the functioning of the world market, which is a classic and increasingly weighty domain of governments and central banks, becomes more and more important.

These few remarks may suffice to substantiate the continued need for intervention. What is more, the interventionist state has neither fallen into disrepute among the population nor among the organizations of capital and labour. The same can be claimed for the welfare state which still meets with almost unanimous approval. In view of the far-reaching socio-structural changes which can be expected in the future, the maintenance of the welfare state is indispensable to check the trend towards societal split – as a result of the tensions between the winners and losers – by distributing work and income in a socially balanced way. The general problem is to transform the total wealth created by the application of new technology into welfare for all. Such a process, however, presupposes not less, but more interaction between the state, the parties and the interest organizations of capital and labour which must be capable of co-operation.

The conservative battle-cry for less corporatism may be suspected of simply promoting another, and no less intensive, form of corporatism. The target is a close co-operation between the state, big business and the banks, to be secured by exclusion strategies. The core of conservative ideologies and policies involves strategies of labour exclusion which not only affect trade union organization but also large parts of the population. The chances of succeeding with such a policy vary from nation to nation but there are many indications that exclusion strategies cannot be pursued at will, i.e. neither as extensively nor as long as desired, and that they can also not be glossed over ideologically. Either conservative policies produce their own opposition as a result of the emergence of political and social counterforces, or they involve the long-term risk that a nation lags behind in the technological race which at all times is also a competition for societal models. Moreover, a new long upturn cycle – even if interrupted by interim recessions – will improve the position of the trade unions as the traditional pioneers of the welfare and social state. Their permanent exclusion from the plant, economic and political bargaining markets will be less likely the more they are capable of representing a broader spectrum of social interests by including new social groups.

Trade Unions between Exclusion and Participation

The obvious question, which can be answered only tentatively, is how the trade unions can survive the transition phase. There is no doubt that their organization and capacity for action have been impaired. However, it would be rash simply to extrapolate current trends and predict continued decline and political exclusion. It is probable that they will hold out till the emergence of a new long wave and will even use this period to ensure that they remain a recognized socio-economic-political partner. That trade unions are not at the end of their tether as intermediary co-operative organizations presupposes: (a) that they are capable of learning processes and can thus initiate renewal through change; (b) that the general political and economic conditions will improve; and (c) that the social tensions which are concomitants of technological change will maintain the classical fields of action for the unions and will add new ones.

The loss of trade union members which can be observed in many areas is primarily a consequence of the fall in employment in the industrial sector and the limited expansion of labour markets in other sectors of the economy. It is unemployment which has caused the decrease in membership and unionization, not – using Hirschman's terminology (1970) – the exit of existing members and loss of loyalty among potential members. Because industrialization and rationalization of services and private and public administration produce a 'proletarian' work environment in these sectors, recruitment opportunities will open up for unions in areas which have hitherto only had a low level of membership. Representation of the interests of the unemployed, of those who are not employed on a regular basis and those who are looking for work and are faced with restricted admission to the labour market, are also opportunities to recruit new members and to extend the trade union monopoly of representation to underprivileged sections of the population. Therefore the fact that with an increasing diffusion of new technology, the productivist core of the economy no longer coincides with the trade union membership employed in the classical industrial sector, does not at all imply political isolation of the trade unions. Being able to exploit such opportunities requires a capacity on the part of the unions to represent, in a much more differentiated way than until now, the heterogeneous social interests of full-timers, part-timers and those without employment, and to resist the conservative temptation to confine themselves to the role of acting narrowly on behalf of the interests of the workforces in the new productivist core sector.

The appeal that conservative strategies can have for the trade unions, the readiness to identify with the winners and to get actively involved in the establishment of a new variant of 'two nations' (Jessop *et al.*, 1984), should not be underestimated. This, however, would mean the end of a trade union movement which is capable of intervention and social participation. Such a development is not inescapable: within the trade union movement itself it is increasingly realized that a merely particularistic interest representation means the loss of the status which had been achieved in the post-war era. The trade unions can start from the fact that their traditional struggle for participation in national wealth – whether in the form of increased income, the reduction of working time, improved working conditions, social welfare benefits or the maintenance of acquired collective protective rights – i.e. their fight for a socially balanced and basically non-exclusive distribution of work and welfare – will also in future meet with widespread approval. In order to counteract the possible division of society into a sector which benefits from the 'new wealth' and another which is characterized by the 'new poverty', it is necessary for the trade unions to turn the further disparity of interests of social groups and subject areas more than before into a crucial issue. Social equality is not necessarily undermined by increased differentiation; the tendency towards equalization and levelling in trade union policy has to be balanced by a stronger consideration of individual, group-specific, and workplace-oriented interests.

The notion of flexibility, a conservative ideological catchword for breaking up collective rights by deregulation, has a strong appeal for many because it seems to promise liberation from rigid work and social practices. The trade unions need to adopt and reinterpret it in order to demonstrate that individual freedom and collective regulation are not opposed but complementary. The obverse of a representation policy in favour of more heterogeneity is, however, that the still important function of the trade unions to focus and aggregate interests becomes more difficult. As a perspective for society as a whole, unions must propagate the humane organization of a world increasingly dominated by technology; and within their own organizations they must develop structures which make it possible to reconcile the particular and the overall interests of their members.

As in the present transition phase the economic functions of the plant and enterprise increase, the regulation of collective relations at the micro level becomes more and more important. In this case the trade unions face two dangers. On the one hand, as a result of tendencies towards 'Americanization' they can be ousted from the bargaining market at plant level; on the other hand there is the danger of the integration of the workforces and their representatives, and their identification with the objectives of the company through a new plant-oriented micro-corporatism which is based on a 'Japanization' of social relations and on employee involvement, and in particular information and consultation bodies below the bargaining levels. In the long run such employers' strategies are aimed at superseding collectively negotiated industrial relations by individually agreed employee relations. Since the introduction of multi-purpose and flexible technology is accompanied by a high demand for responsible and flexible labour, so that 'high-trust systems' (Fox, 1974) are emerging, a struggle between management and trade unions for predominance in industrial and social relations at the workplace is about to occur. Trade unions can only withstand this trial of strength if they are provided with central structures which enable them to exert an influence on social relations at plant level. Only bi-polar organizational structures, permitting a thorough social and political representation of interests from the micro to the macro level, can safeguard the unions against dangers inherent in the tendency – which is thought to be inevitable not only by conservatives – of allowing workplace representatives more latitude while acting merely 'as an umbrella organization giving general guidelines' (Späth, 1985: 38). More than before, their organization and capacity for mobilization and bargaining power will depend on the balance between decentralized and centralized principles of organization.

Development in the Western European countries is not at all homogeneous. Rather there is a polarization between models of labour exclusion and of participatory interaction, involving a contrast between a British power model and a continental model of participation. Such contrary developments are not only a result of a far more ambitious and more rigorously imposed British variant of conservative policy, in comparison with the countries on the Continent, but also relate to the problem of a balanced trade union structure. The fact that in Great Britain it has been possible to promote a model of society based on strategies of labour exclusion and mini-corporatism can be attributed to a large extent to a predominance of plant-centred trade union structures (Jacobi and Kastendiek, 1985). The lack of industry-wide trade unions with centralized powers and responsibilities has – more than in other countries – led to inter- and intra-trade union competition, and has made it easier for the government to exclude trade unions by political and legislative measures and to discipline and integrate the workforces and their representatives at plant level. In contrast, the participatory model of the post-war period in a number of Continental European countries – even if weakened – does not seem to be impaired indefinitely, so that the trade unions can maintain their presence in the enterprise, economic and political markets if they successfully achieve renewal through change.

Conclusions

The aim of this chapter has been to indicate some plausible counter-arguments and counter-tendencies, in order to show that new technological paradigms by no means imply a long-term recession and the demise of the trade unions and of welfare democracy.

The advantages inherent in the new technology are so enormous that it would be completely wrong – and also not very realistic – to attempt to thwart its development and application. This does not mean that everything should be done that is technically feasible. Only if it is possible to make a sensible choice, can the new technology become the 'key to utopia'. 'The vision of a world without poverty and freed from the toil of physical work' (King, 1982: 43–4) will be technically feasible in the future. Whether it is politically feasible remains uncertain. Controlling the destructive potential of new technologies presupposes a 'creative partnership' between the state, society, trade unions and science (King, 1982: 44). The problem is to pursue an 'intelligent interventionism' on a national and international basis which benefits all. If this can be achieved the trade unions have a good chance of influencing the application of future technology for the benefit of humanity. The countries with a participatory-consensual model of society are likely to be the pioneers of a future society.

Quite a few readers may feel that the perspective of invervention and interaction within the framework of capitalist conditions (in economic terms, a form of super-Keynesianism) – is too restricted. But regardless of whether this is an inexcusable moderation on the part of the author, or whether socialist perspectives are beyond all feasible reality, the solution of the 'second best' seems a worthwhile objective.

References

Anders, G. 1984. 'Brecht konnte mich nicht riechen'. *Die Zeit*. Vol. 13, no. 22. March.

Berger, J. 1983. 'Die Wiederkehr der Vollbeschäftigungslücke – Entwicklungslinien des wohlfahrtsstaatlichen Kapitalismus'. *Krise der Arbeitsgesellschaft*. Ed. J. Matthes. Frankfurt/New York: Campus.

Blackburn, P. 1985. *Technology, Economic Growth and the Labour Process*. London/Basingstoke: Macmillan.

Borner, S. 1981. 'Die Internationalisierung der Industrie'. *Kyklos*. Vol. 34.

Coombs, R. W. 1984. 'Die Verbreitung von Mechanisierungstechniken und Theorien der Langen Wellen'. *Prokla*. Vol. 57, 79–98.

Dosi, G. 1983. 'Technological Paradigms and Technological Trajectories. The Determinants and Directions of Technical Change and the Transformation of the Economy'. *Long Waves in the World Economy*. Ed. C. Freeman.

Fox, A. 1974. *Beyond Contract: Work Power and Trust Relations*. London.

Freeman, C. (ed.) 1983. *Long Waves in the World Economy*. London.

Galtung, J. 1983. 'Die globale Verteilung von Wachstum und Stagnation'. *Vor uns die goldenen neunziger Jahre?* Ed. M. Jänicke.

Hirschman, A. 1970. *Exit, Voice and Loyalty*. Cambridge, Mass: Harvard University Press.

Huber, J. 1985. 'Modell und Theorie der langen Wellen'. *Vors uns die goldenen neunziger Jahre?* Ed. M. Jänicke.

Jacobi, O. and H. Kastendiek. (ed.) 1985. *Staat und industrielle Beziehungen in Großbritannien*. Frankfurt/New York.

Jänicke, M. 1985. 'Langfristige Wachstumsperspektiven der westlichen Industrieländer. *Vor uns die goldenen neunziger Jahre?* Ed. M. Jänicke. München/Zürich.

Jessop, B. *et al*. 1984. 'Authoritarian Populism, Two Nations and Thatcherism'. *New Left Review*. September/October.

Kahn, H. 1983. *Der kommende Boom*. Bern/München.

King, A. 1982. 'Einleitung: Eine neue industrielle Revolution oder bloß eine neue Technologie?' *Auf Gedeih und Verderb – Mikroelektronik und Gesellschaft*. Ed. G. Friedrich and G. A. Schaff. Wien.

Kleinknecht, A. 1984. 'Innovationsschübe und Lange Wellen: Was bringen "Neo-Schumpeterianische" Kriseninterpretationen?'. *Prokla*. Vol. 57, 55–78.

Landesregierung Baden-Württemberg. 1983. *Zukunftsperspektiven gesellschaftlicher Entwicklungen*. Stuttgart.

Lutz, B. 1984. *Der kurze Traum immerwährender Prosperität*. Frankfurt/New York.

Matzner, E. 1982. *Der Wohlfahrtsstaat von morgen*. Frankfurt/New York.

Meißner, W. and K. Zinn. 1984. *Der neue Wohlstand*. München.

Offe, C. 1983. 'Competitive Party Democracy and the Keynesian Welfare State: Some Reflections upon their Historical Limits'. *The State, Class and the Recession*. Ed. S. Clegg et al. London/New York.

Olson, M. 1982. *The Rise and Decline of Nations*. New Haven/London.

Piatier, A. 1983. 'Innovation, information and long-term growth'. *Long Waves in the World Economy*. Ed. C. Freeman.

Skidelsky, R. 1977. *The End of the Keynesian Era*. London/Basingstoke: Macmillan.

Späth, L. 1985. *Wende in die Zukunft*. Hamburg.

Wallerstein, I. 1983. 'Die Zukunft der Weltökonomie'. *Perspektiven des Weltsystems*. Ed. J. Blaschke. Frankfurt/New York.

12

Technological Change and Unions

Greg Bamber

In spite of all the writing on recent technological changes (for bibliographies, see Ford *et al.*, 1981; Wooden and Kreigler, 1985; Work Research Unit, 1985), it is also worth examining some earlier experiences, not least because contemporary advocates of technological change often accuse unions of nineteenth-century Luddism. The Luddites emerged between 1811 and 1813 in the Midlands and North of England. In night-time raids they smashed textile machines and wrecked whole factories. In so far as they had a 'leader', he was sometimes referred to as Nedd Ludd or General Ludd.

There are few available written records about the Luddites' objectives. Hence we are not sure to what extent they were opposing technological change as it would displace hand-workers or dilute their skills, or trying to persuade the masters who owned the machines to award higher pay and better conditions to their men (Pelling, 1976: 28–9; O'Brien and Engerman, 1981: 161). It is worth recalling that the Luddites were not organized into unions, which were then illegal and suppressed (Thompson, 1968: 537ff). If they had been able to channel their grievances through a union, their forms of protest might have been rather different. Incidentally, Luddism was not uniquely British; there were anti-machine protests in several other countries as they began to industrialize.[1]

The early British unions established mutual insurance schemes, which could, among other things, help to compensate those displaced by technological change. Some unions tried to regulate technological change unilaterally. Increasingly, unions engaged in collective bargaining, as a way of regulating technological change bilaterally (through negotiations with employers). Unions also sought legal enactment, for instance, to influence minimum pay levels and health and safety issues.

Such state intervention could be a form of trilateral regulation (involving three parties: unions, employers and the state). These union methods of job regulation generally involved compromise, rather than violent opposition. None the less, the mid-nineteenth century conventional wisdom was that unions oppose technological change. Even though the 1894 Royal Commission examined the 'rules and regulations' of hundreds of unions, however, 'in none of them did it discover any trace of antagonism to invention or improvement' (Webb and Webb, 1913: 393). By contrast, where unionism had become reasonably established, it concluded that:

> the old attempt of the handicraftsman to exclude the machine has been abandoned. Far from refusing to work the new processes . . . trade unionists . . . claim, for the operatives already working at the trade, a preferential right to acquire the new dexterity and perform the new service (1913: 411).

Contemporary researchers might criticize the research methods of the Royal Commission and the Webbs for placing too much weight on formal documents rather than other forms of research, yet their conclusions are supported by several recent studies, which use quite different methods.

In the USA, for example, Weikle and Wheeler surveyed the attitudes of local union leaders to technological change. The average response was 'a rather mild form of encouragement', but these trade unionists wanted to influence the kind of changes which take place (1984: 100–6). In a comparative study, Northcott *et al.* (1985) asked the technical director (or equivalent) in manufacturing establishments about a wide range of problems in the use of microelectronics. They found that opposition from the shopfloor or unions was not seen as a major problem. It was identified by only 16 per cent of manufacturing establishments in France, 14 per cent in West Germany and by only 7 per cent in Britain. In these countries, union opposition was seen as a much less important obstacle than such other issues as the general economic situation, lack of people with microelectronic skills, and high costs of development. Interestingly, in Britain, opposition from top management was seen as a problem 'by 5 per cent of the establishments, slightly less often than trade unions, but the figure may under-represent the extent of the difficulty since a proportion of the respondents would probably regard themselves as top management and would be unlikely to see themselves as a major obstacle' (Northcott *et al.*, 1985: 38).

Union Policies

Large unions are generally less likely to oppose technological change than are individuals or particular groups of workers who fear that their own position in the labour process will be adversely affected. Therefore it is important to distinguish the position of individuals or workgroups, from that of unions as collective organizations (cf. Hyman and Fryer, 1975). Most union leaders publicly welcome investment in new technology. From the vantage point of a national union head office, it may be relatively easy for a union leader to adopt such a view, in contrast with a worker or workgroup which is liable to be directly displaced by a particular change. As negotiators, however, union leaders aim to influence how technological change is introduced. They do not want to stop it, but rather to control it, whether unilaterally, bilaterally or trilaterally.

There is much more scope for choice about the use of microelectronic technology than with earlier mechanical technology. (Microelectronic devices are smaller, less dependent on energy, more easily transported and more flexible.) Contemporary unions have more professional expertise than their predecessors. Nevertheless, the role of unions is constrained, in so far as they are in a generally unfavourable economic and political context in the 1980s and unions rarely *initiate* the introduction of new technology, but *react* to employer-inspired initiatives.

Since the late 1970s, many unions have reconsidered their policies in relation to the increasingly widespread use of microelectronic technology. Such policies have often been published or codified internally by national unions and centres, as well as by various international union organizations.[2]

An 'Ideal-type' Union Technology Policy

Although their policies differ, one way of explaining them is to construct an 'ideal type'[3] union policy. This includes a series of procedural and substantive objectives.

Procedural Objectives

Consultation. Consultation should begin by employers disclosing full information about the proposed change, at the contemplative stage in the decision-making process, so that unions can genuinely influence the choice of technologies and how they are used, rather than merely influencing the minor details of implementation.

Union expertise. Unions should retain their own independent technical specialists, who can appraise employers' proposals. These may be external consultants and/or internal 'technology stewards', who should be trained and have access to the employers' technical specifications and to other data. There should be extensive education of union officials and members about the technical and social issues associated with technological change.

Data protection. Information technology should not be used to invade workers' privacy. Employers should not use it to control an individual worker's performance, unless agreed in advance. Unions should have joint control of any data which are collected, and decide how they are used and who can have access. (Individuals should have access to any data which relate to them.)

Joint reviews. Unions should be involved in regular reviews of any technological change, to ensure that the agreed policies are being followed and so that any unanticipated consequences can be dealt with.

Substantive Objectives

Job security. Technological change should not be used to reduce the number of jobs. If new technology is used to increase productivity, the output or service should be increased, rather than the volume of labour decreased. If such a decrease is unavoidable, people should be redeployed, with no loss of pay or conditions. Failing that, they should be encouraged to volunteer for redundancy or early retirement, so that the decrease is achieved voluntarily. Such 'natural wastage' is not welcomed, but it is preferable to 'compulsory redundancy'.

Redundancy. If compulsory redundancy becomes unavoidable, the selection criteria, compensation, and length of notice should be negotiated in advance. Those affected should be counselled and helped to find alternative employment.

Retraining and reskilling. Whenever possible, those people displaced should be retrained to work with new technologies, or elsewhere. Thus, technological change should not lead to deskilling; it should be used to create opportunities for reskilling. Employers should invest in training their employees (preferably during working hours).

Working hours. Technological change should be used to offer workers more choice about when they work and a reduction in their total working hours, for example, by introducing longer holidays, a shorter working week, earlier retirement, sabbatical leave, and the elimination of systematic overtime. It should not necessarily be used to increase the numbers of people who have to work shifts.

Pay. Employers should pay extra money to the people who learn new skills. Even if no new skills are required, technological change should be accompanied by a pay increase, rather than a reduction.

Job design.   The opportunity should also be taken to *improve* the physical and psychological quality of the working environment; for example, workers should be able to control their own pace and quality of work, which should not be too repetitive or fragmented.

Health and safety.   Similarly, workplace health and safety should be enhanced, rather than the reverse. New hazards and stresses which may be associated with new technology should be eliminated, before innovations are commissioned.

Equal opportunities.   Technological change should not precipitate increased polarization between a highly-skilled, well-paid minority and a deskilled, low-paid majority. People should have equal opportunities to be appointed to the best jobs, irrespective of their gender, race, religion or family background.

This 'ideal-type' policy represents a broad generalization and is not always applicable. There has been a growing union concern with such issues as health and safety, equal opportunities and job design; but these concerns have not arisen uniquely in the context of technological change. Little of this 'ideal-type' union policy was formulated specifically to confront current technological changes. Rather, most elements of the policy are adaptations of earlier policies. Perceived threats associated with microelectronic innovation have provided union policy-makers with opportunities to consolidate their strategies and tactics. Coping with technological change has generally been a newer experience for predominantly non-manual unions than for predominantly manual or craft unions. The use of much machinery in offices is more recent than in many fields of manual employment.

The formulation of strategies and tactics in any particular case involves all the usual internal debates and negotiation processes among union members and officials (cf. the 'attitudinal structuring' and 'intraorganizational bargaining' analysed by Walton and McKersie, 1965). The outcome usually reflects the relative power of different factions or interest groups and their particular priorities.

Union Responses

When union negotiators establish their priorities, in exchange for a high level of pay, for example, some workers may be willing to tolerate a repetitive job design or unsafe working conditions. Hence, it is important to distinguish between union policies and union responses in practice. Policies may reflect long-held ideological orientations. Responses are influenced by such policies, but also by the immediate context, over which employers (and governments) may have more influence than union leaders.

We can begin to classify union responses into the following five categories: (1) participative involvement; (2) negotiated trade-offs; (3) unconditional acceptance; (4) reluctant acquiescence; (5) complete resistance.[4]

1. *Participative involvement* occurs where unions positively welcome technological change, and have a real input into the fundamental decisions about choices and design at the formative stage. Such behaviour would seem to follow if our 'ideal-type' policy were fully implemented in spirit by all concerned in the change: managers, union representatives and individual employees. This rarely happens. In practice, as in the following four categories, there is little or no union or workgroup involvement in making the formative decisions, which are made more or less unilaterally by the employers' engineers and managers.

2. *Negotiated trade-offs* means that unions accept a technological change in exchange for certain trade-offs, which usually relate to how the change is implemented. This may involve a *quid pro quo* between pay and conditions, for example.

3. *Unconditional acceptance* also means that employers make the decisions unilaterally, but that they may then successfully 'sell' the change directly to employees and their unions. In spite of their leaders' policies, some workgroups may not want to participate in decision-making, but accept that 'employers should manage'. This position may be found with unskilled workers and in new establishments, where there is no union with which to bargain or consult.

4. *Reluctant acquiescence* again means that employers make the decision, but present it to unions on a 'take it or leave it basis', implying that the stark alternative is unemployment. This has increasingly been the case against the background of economic recession, when unions may not have enough power to oppose decisions successfully.

5. *Complete resistance* by unions is rare and usually relatively short-lived, but may occur if the union leaders and members believe that the change will have unmitigated deleterious consequences for them and that these cannot be sufficiently ameliorated by negotiating or consulting with management. This position also implies that the union can exert some power in relation to the employers.

TABLE 12.1

Approaches to the Joint Regulation of Technological Change

Country	Approximate Unionization Rate (%)	Laws Used	Collective Agreements		
			National Framework Level	Sectoral Level	Company or Plant Level
West Germany	40	Works Constitution Act 1972; Works Safety Act 1973 and regulation on work with VDUs 1981	None	Job protection agreements in metalworking, textiles, footwear, leather, paper processing, printing	More than 100 agreements concluded
Norway	45	Working Environment Act 1977 and regulation on work with VDUs 1982	Agreement on computer-based systems 1975	Banking	Most of industry and services covered by local agreements
Sweden	73	Working Environment Act 1978 and regulation on work with VDUs 1981; Co-determination Act 1977	Work environment agreement 1976	Technology agreement in the printing sector, co-determination agreements in central government, local government, private sector	Use of legislative rights

Italy	43	Statute of Workers Rights 1970 Health and Safety Act 1978	None	Clauses included in sectoral agreements on metalworking	Clauses included in several agreements e.g. Fiat, Olivetti, Alfa Romeo
USA	<20	Occupational Safety and Health Act 1970	None		Technical change clauses in collective agreements
Canada	30	Recommendations for legal rights	None		Limited clauses in existing agreements
UK	50	Health and Safety at Work Act 1975	None	Parts of public sector	More than 100 agreements concluded
Australia	55	None	Decisions of Federal Conciliation & Arbitration Comm. (e.g. job security)	Telecommunications	Agreements in parts of printing and public sector

Source: Adapted from Commonwealth Secretariat (1985: 94–5), after Lansbury (1986: 17).

International Differences

Table 12.1 summarizes national contexts for regulating the use of new technology at the workplace. It can be argued (Bamber and Lansbury, 1987: 23–4), that in countries with adversarial traditions of industrial relations (most English-speaking countries), unions are less likely to co-operate with technological change than their counterparts in countries with recent traditions of social partnership in industrial relations (West Germany and the Scandinavian countries). To a considerable extent, current differences in union behaviour reflect contrasting legacies of employers' attitudes (cf. Fox, 1978). For example, American and British unions have traditionally placed more emphasis on bargaining after decisions have been made, rather than on participation in making decisions, in contrast with many of their German and Scandinavian counterparts, which face more paternalistic employers.

Even in countries with adversarial traditions, in the context of heightened international competition and the introduction of new technology, many managers are aiming to introduce quasi-paternalistic programmes which may be called employee involvement (EI) or quality circles (QCs). With their tradition of 'business unionism' American trade unionists have more often been persuaded to accept such innovations in human resource management than their British counterparts, many of whom have a stronger ideological orientation. Nevertheless, at greenfield sites, in particular, some unions in Britain are making controversial new forms of collective agreement, which welcome new technology, establish flexible working practices, often include types of EI, and pendulum arbitration (known in the USA as final-offer arbitration). Some such agreements have been called 'no strike deals', especially those between multinationals (mostly Japanese) and the EETPU. But in their concern to reverse declines in membership, several other unions are also making such agreements.

Some British trade unionists are reconsidering their traditions and concluding that collective bargaining has not always proved an effective means of regulating technological change. Union policy-makers are seeking to supplement collective bargaining, for instance, by government action and legal enactment to stimulate economic growth, foster some forms of industrial democracy and to limit overtime working. Moreover, at company level, there have been some interesting workers' plans to initiate alternative technologies (cf. Wainwright and Elliott, 1982).

Scandinavian unions have tended to put more emphasis on job design than unions in most other countries, but have generally not adopted the tactic of aiming to cut working hours. However, international differences in union strategies are probably less significant than other differences, for example, between different types of occupations, unions and sectors.

Differences Between Sectors

In many countries, craft printing-workers' unions have tended to resist technological change, while unions in the electronics industry have tended to co-operate with it. How can we explain such differences? Following Slichter *et al.* (1960), we can begin to predict union responses by posing and answering four questions. First, is it a craft or industrial union? (The former generally oppose change which destroys the basis of their craft.) Second, what is the state of the product market? (Unions are more likely to accept change where there is a high degree of competition and an expanding market.) Third, what is the type of technological change (in terms of: (i) the number of jobs affected; (ii) the effect on the degree of skills required; (iii) the effect on the kind of skills required)? Fourth, at what stage is the innovation? (Union opposition is most likely at an early stage, when there is still considerable uncertainty about the change.)

The *economic environment* shapes union responses, but can have two contradictory influences. On the one hand, unions generally have more power with which to oppose change where there is a tight labour market, than where there is a slack one (cf. Hunter *et al.*, 1970). Yet, on the other hand, where the labour market is tight, perhaps there is less motive for unions to resist change, given that it should be easier for displaced workers to find alternative employment. Where the labour market is slack, there is likely to be more motive for union resistance, but it is generally less effective.

A study of the transport industry suggests that the following factors incline a union to accept a change: (a) a broad membership base; (b) large size; (c) political security of the union and leadership; (d) absence of inter-union rivalry; (e) a concentration of power within the union; (f) employer unity; and (g) incentive payments which reward workers for achieving higher productivity following technological change (Levinson *et al.*, 1971). Another American study finds that union leaders are most likely to accept change if they perceive that: (a) only a small proportion of the jobs in an affected unit would be lost; (b) the change is inevitable; (c) there could be a *quid pro quo* for the lost jobs (McLaughlin, 1979). Such factors could be generalized to other countries.

Challenges for Unions

Turning from their current policies and responses, what further challenges are unions facing?

Union Organization

The formal organization of a union is often defined in a rulebook, which originated when the union (or its predecessor) was first formed (perhaps in the nineteenth century). Unions could now be organized rather differently, to take advantage of modern communications and computerization. It is difficult to change such a traditional democratic organization, however, particularly if its current leaders have risen to power through it.

Few unions had either the expertise or the available spare cash themselves to pioneer with advanced communications and computer systems. Hence, unions have tended to be slow to make much use of such microelectronic technology, in contrast to some businesses. Nevertheless, in the late 1980s, many unions are themselves introducing technological change. Their membership records and financial control can often be administered more efficiently by computer than by older manual methods. Furthermore, some unions are using new technologies to provide services to their members and officials. Some are trying to improve their research methods, by using electronic data processing to analyse recruitment patterns and prospects, employers' finances, and industrial relations awards, agreements and contracts.

Some unions are considering the use of information technology to improve inter- and intra-union communication. Union officials, researchers, stewards and rank-and-file members may wish to be able to communicate with each other about current issues much more quickly than by written circular, mass meetings, conferences, and postal ballots (cf. Warner, 1984).

Structural Change

In most countries, union boundaries are usually defined either in industrial or occupational terms. In the face of rapid technological change, certain industries and occupations are declining or even becoming obsolete. In other cases, former distinctions between industries and occupations are diminishing and sometimes disappearing. These changes pose great challenges for unions, especially for those whose boundaries are defined in terms of a declining occupation. Many such unions have looked to broaden their coverage and for 'appropriate' partners to merge with. The criteria of appropriateness may be defined with regard to technology. In practice, however, most unions are more concerned to look for a partner which has compatible forms of government, will strengthen their bargaining power and extend their membership. Also, union officials often have their own reasons for seeking mergers, which may be motivated by political, personal, and parochial interests (Mansfield, 1981; Bamber, 1986).

Unions generally find that it is easier to recruit a high density of membership in the public sector and the older heavy industries, especially in the larger establishments, where there is a high concentration of potential members. Partly as a reflection of technological change, however, a declining percentage of the workforce is employed in heavy industry. Most establishments are becoming more capital-intensive. There are relatively more 'knowledge' workers and fewer manual workers. Moreover, the number of people employed in the service sector is growing, relative to manufacturing. Toffler (1983) calls this a shift from the second to the third wave. He foresees a continuing move away from mass production towards a 'de-massified' economy. Corporations are increasingly using sophisticated information technology to decentralize and even to fragment the organization of work. People are being employed in smaller units, rather than in large factories and offices. There is also a revival of latter-day homeworking by subcontractors, who may communicate by computer technology. But homeworkers are difficult for unions to recruit and thence to mobilize.

Such structural changes represent profound challenges for the union movement. These changes have exacerbated the relative decline of union density in Japan, the USA and the UK in the early 1980s.[5]

Corporate Decision-making
Although corporations may be decentralizing the organization of work and the *implementation* of human resourcing practices, they mostly continue to determine the *corporate strategies* and major investments in technological change at a central level. In the UK, Canada and the USA, for instance, where most private-sector collective bargaining is devolved to a local level, unions experience great difficulty in influencing the conception and implementation of technological change. There is often a mismatch between the most important level of company decision-making and the level at which unions bargain. None the less, even in West Germany and other countries where unions have boardroom representation, they still complain that they are unable to exert much real influence over technological change (Bamber and Lansbury, 1986). This complaint is not only in relation to foreign-owned multinational corporations. Even in locally-owned firms, unions are also concerned that they have little access to the committees which shape corporate strategies and that design criteria may be established by technical experts, or in other locally-owned companies to which they have no access.

If we accept that employee participation in decision-making seems likely to facilitate better decisions and to increase employee commitment, there is a considerable challenge for managers too: to try to increase the opportunities for employees and their unions to be involved in a genuine way.

Conclusion

Technological change does not generally raise fundamentally new issues for unions. Nor is it a discrete issue which can be dealt with separately from most other issues associated with the employment relationship. Most elements of our 'ideal-type' policy were on union agendas long before the appearance of microelectronics. However, the rapidity and extent of current changes add up to a great challenge for unions, especially against the background of high unemployment and the changing international division of labour.

This challenge relates to and further complicates most of the existing union priorities (Moore and Levie, 1985). Hence discussion of these issues should not try to focus on new technology and unions in a vacuum, but should recognize the historical, economic, political and social context of industrial relations.

Notes

This chapter draws on a survey of the main international trade union secretariats based in Western Europe. It also draws on case studies conducted in Australia, Britain and West Germany. The author is grateful to all who have helped in these endeavours and to several colleagues for helpful comments on a version of the paper given in 1986 to the Seventh World Congress of the International Industrial Relations Association (Switzerland).

1. For an account of a Luddite attack in Britain, see Charlotte Brontë's *Shirley*, London: Octopus, 1849. For a vivid description of anti-machine protests in France, see Stearns (1972: 25ff).
2. For details of union policies in various countries: see, on Australia: Markey (1983); on Britain: Williams and Steward (1985); on Sweden: Anderman (1967) and Swedish Trade Union Confederation (LO) (1983); on Norway: Keul (1983); and on various other countries, see Bamber and Lansbury (1983); European Trade Union Institute (1982). For examples of some international union policies, see Watts and Pitous (1981); FIET (1983); Rush and Hoffman (1983); ICFTU (1984); EMF (1985).
3. The 'ideal-type' concept is used here in the Weberian sense as an exaggerated abstraction. It simplifies, to illustrate typical union policies on technological change, but the various items are 'more or less present and occasionally absent' in particular cases. Actual policies and practices can be better understood by comparing them with such an ideal type. This does not imply a moral value of 'the ideal'; see Weber (1949). In view of space constraints, this whole discussion focuses on union action *vis à vis* employers. But of course political action is also a crucial component of union policies on technological change.
4. For other approaches to classifying union behaviour, see Slichter *et al.* (1960); Francis and Willman (1980); Evans (1983).
5. Of course there are other reasons for the decline in unionism, such as the

increase in unemployment, particularly in the UK. Also, American and British unions blame the anti-union policies of President Reagan and Prime Minister Thatcher for some of the continuing decline in this period.

References

Anderman, S. D. (ed. and tr.) 1967. *Trade Union and Technological Change*. A Research Report Submitted to the 1966 Congress of Landorganisationen I Sverige (LO). London: Allen & Unwin.

Bamber, G. J. 1986. *Militant Managers? Managerial Unionism and Industrial Relations*. Aldershot: Gower.

Bamber, G. J., and R. D. Lansbury (ed.). 1983. 'Technological Change and Industrial Relations: An International Symposium', Special Issue of the *Bulletin of Comparative Labour Relations*. Vol. 12.

Bamber, G. J., and R. D. Lansbury. 1986. 'Codetermination and Technological Change in the West German Automobile Industry'. *New Technology, Work and Employment*. Vol. 1, no. 2.

Bamber, G. J., and R. D. Lansbury (ed.). 1987. *International and Comparative Industrial Relations: A Study of Developed Market Economies*. London: Allen & Unwin.

Commonwealth Secretariat. 1985. *Technological Change: Enhancing the Benefits*. Report by a Commonwealth Working Group. Vol. 2. London: Commonwealth Secretariat Publications.

EMF. 1985. *EMF Basic Demands with regard to Technological Change and a Changing Society*. Brussels: European Metalworkers' Federation (EMF).

ETUI. 1982. *Negotiating Technical Change*. Brussels: European Trade Union Institute (ETUI).

Evans, J. 1983. *Technological Change: A Trade Union Perspective*. Address to International Graphical Federation Congress, Paris: IGF.

FIET. 1983. *Model Technology Agreement*. Geneva: International Federation of Commercial, Clerical, Professional and Technical Employees (FIET).

Ford, B., M. Coffey, and D. Dunphy. 1981. *Technology and the Workforce*. Sydney: Technology Research Unit.

Fox, A. 1978. 'Corporatism and Industrial Democracy: The Social Origins of Present Forms and Methods in Britain and Germany'. *Industrial Democracy: International Views*. Coventry: Industrial Relations Research Unit, University of Warwick.

Francis, A., and P. Willman. 1980. 'Microprocessors: Impact and Response'. *Personnel Review*. Vol. 9, no. 2. Spring.

Hunter, L. C. *et al.* 1970. *Labour Problems of Technological Change*. London: Allen & Unwin.

Hyman, R., and R. H. Fryer. 1975. 'Trade Unions: Sociology and Political Economy'. *Processing People: Cases in Organisational Behaviour*. Ed. J. B. McKinlay. London: Holt, Rinehart and Winston.

ICFTU. 1984. 'Technology, Growth and Development'. *ICFTU 1984 World Economic Review: The Reality of Interdependence*. Brussels: International Confederation of Free Trade Unions (ICFTU).

Keul, V. 1983. 'The Trade Union Movement, Research and Data Technology:

an Account of Three Research Assignments for Trade Unions'. *Computeriz-ation of Working Life*. Ed. E. Fossum. Chichester: Ellis Horwood.

Lansbury, R. D. 1986. 'Technological Change and Industrial Relations: General Report'. *IIRA 7th World Congress Proceedings*. Vol. 1, Geneva: International Industrial Relations Association.

Levinson, H. M., C. M. Rehmus, J. P. Goldberg, and N. L. Kahn. 1971. *Collective Bargaining and Technological Change in American Transportation*. Evanston: Northwestern University.

McLaughlin, D. B. 1979. *The Impact of Labor Unions on the Rate and Direction of Technological Innovation*. Washington: National Technical Information Service (PB - 295 084).

Mansfield, B. 1981. 'Technological Change and the Trade Unions'. *Labor Essays 1981*. Eds. G. Evans *et al*. Richmond, Vic.: Drummond.

Markey, R. 1983. *The Trade Union Response to Technological Change in Australia*. Sydney: Industrial Relations Research Centre, University of New South Wales.

Moore, R., and H. Levie. 1985. 'New Technology and the Unions'. *The Information Technology Revolution*. Ed. T. Forester. Oxford: Basil Blackwell.

Northcott, J. *et al*. 1985. *Microelectronics in Industry: An International Comparison*. London: Policy Studies Institute.

O'Brien, P. K., and S. L. Engerman. 1981. 'Changes in Income and its Distribution During the Industrial Revolution'. *The Economic History of Britain Since 1700*. Ed. R. Floud and D. McCloskey. Cambridge: Cambridge University Press.

Pelling, H. 1976. *A History of British Trade Unionism*. Harmondsworth: Penguin.

Royal Commission on Labour 1891–94. 1894. London: HMSO (C. 6708).

Rush, H., and K. Hoffman. 1983. 'From Needles and Pins to Microelectronics: The Impact of Technical Change in the Garment Industry'. International Textile, Garment and Leather Workers' Federation, Paper to ITGLWF meeting on Technological Change in the Clothing Industry.

Slichter, S. H. *et al*. 1960. *The Impact of Collective Bargaining on Management*. Washington: Brookings.

Stearns, P. N. (ed.) 1972. *The Impact of the Industrial Revolution*. New York: Prentice-Hall.

Thompson, E. P. 1968. *The Making of the English Working Class*. Harmondsworth: Penguin.

Toffler, A. 1983. *Previews and Premises*. London: Pan.

Wainwright, H., and D. Elliott. 1982. *The Lucas Plan: A New Trade Unionism in the Making?* London: Allison and Busby.

Walton, R. E., and R. B. McKersie. 1965. *A Behavioral Theory of Labor Negotiations*. New York: McGraw-Hill.

Warner, M. 1984. 'The Impact of New Technology on Participative Institutions and Employee Involvement'. *The Management Implications of New Information Technology*. Ed. N. Piercey. London: Croom Helm.

Watts, G. E., and C. Pitous. 1981. *Impact of Technological Changes*. Geneva: Postal, Telegraph and Telephone International (PTTI).

Webb, S., and B. Webb. 1913. *Industrial Democracy*. London: The Authors (most of the book was written in the 1890s).

Weber, M. 1949. *The Methodology of the Social Sciences*. Glencoe, Ill: Free Press.

Weikle, R. D., and H. N. Wheeler. 1984. 'Unions and Technological Change: Attitudes of Local Union Leaders', *Proceedings of the Thirty-Sixth Annual Meeting*. Industrial Relations Research Association Series, Madison: IRRA.

Williams, R., and F. Steward. 1985. 'New Technology Agreements: an Assessment'. *Industrial Relations Journal*. Vol. 16, no. 3.

Wooden, M., and R. Kreigler. 1985. *Technological Change and its Implications for Industrial Relations*. Adelaide: National Institute of Labour Studies Working Paper Series No. 78.

Work Research Unit (WRU). 1985. *New Technology: The Trade Union Response*. Information System Bibliographies No. 24, London: WRU.

13

Policy Debates over Work Reorganization in North American Unions

Harry Katz

Changing conditions of world competition and technological innovation have put pressure on North American unions to redesign collective bargaining and work practices. In some industries and firms, work reorganization and new forms of labour–management interaction have emerged, while in some other cases change has been opposed outright or innovation has stalled. Everywhere, national as well as local unions have begun to debate alternative strategic responses to economic pressure and industrial adjustment.

This chapter argues that much of the debate within the ranks of North American labour involves an argument between two basic policy approaches – a 'co-operatist' and 'militant' strategy. These strategies are distinguished by their views regarding the proper course of: pay (level and method of determination); work reorganization; the roles of workers and unions; and government's role in industrial restructuring. In this chapter, reference is made to the statements of union leaders and union policies to characterize the co-operatist and militant positions and to illustrate their application. The chapter goes on to argue that at the root of this policy debate are different assumptions regarding product demand and technological developments and most importantly, different ideological premises.

While these assumptions have helped to produce starkly contrasting broad union strategies, the practice of union policy has sometimes led to compromise solutions from both camps. The militants sometimes agree to co-operative-like reforms, while the co-operatists have at times militantly opposed certain management demands.

Facing the need to compromise periodically, both the co-operatists and militants find themselves with symmetric problems and in need of a clear plan of action. The problem for the co-operatists is that without a clear vision of either how the various pieces of work reform fit together or how workplace reform could link to wider economic reforms, they find it hard to defend themselves against charges that they are merely passively accepting managerial initiatives. Meanwhile, the militants, lacking a vision of how their resistance to management's demands fits together as a viable strategy that responds to economic pressures, are forced to defend themselves against the charge that their policy is merely an unimaginative defence of the *status quo*. This chapter argues that although the co-operatist and militant positions have been pulled somewhat together by practical responses and each side suffers from the absence of an overall plan, there is no mistaking the fundamental differences that exist in the two positions.

The Co-operatist and Militant Positions

A central difference between the co-operatists and militants is the extent to which each is willing to agree to concessions and their willingness to accept contingent compensation as part of a concessionary package.[1] Confronted by lay-offs or the threat of future employment decline, the co-operatists often are willing to agree either to compensation freezes, slowdowns, or cuts, in order to improve employment prospects. Examples of this approach include the willingness of Lynn Williams, current president of the United Steelworkers (USW), and Don Ephlin, current head of the General Motors (GM) department of the United Autoworkers (UAW), to approve the negotiation of pay concessions.[2]

The co-operatists are willing to agree to pay concessions where they believe that they implicitly (through improvements in competitiveness) or explicitly (through specific contractual job security programmes) lead to more employment. The co-operatists also are willing (and sometimes eager) to agree to compensation procedures that tie future worker pay to company performance. In the case of Ephlin this appears in his willingness to agree to profit-sharing at Ford and GM. Ephlin also negotiated a formula in the agreement covering GM's Saturn project that will set worker pay at 80 per cent of industry standards and allow the rest of pay (conceivably enough to raise Saturn pay above the industry average) to be set as a function of the financial performance of the Saturn Corporation. In the United Steelworkers' recent contracts with the Wheeling-Pittsburgh and LTV steel companies, Williams negotiated formulae which restore pay concessions as a function of future company performance. This is the pattern he followed in the company-by-company negotiations that replaced the traditional national basic steel agreement in 1986.

The militant position, on the other hand, rejects pay concessions, arguing that these are unlikely to produce much in the way of employment expansion, and are an unfair way to cut costs. The militants argue that profits or management salaries instead should be lowered (and conceivably could be lowered enough to restore competitiveness) (Slaughter, 1983). Furthermore, the militants are adamantly opposed to contingent compensation mechanisms that link worker pay to company or work group performance. They believe these pay procedures create divisive splits across workers and see them as a step toward company unionism (in the pejorative American sense).[3]

The co-operatists and militants also differ sharply in their position regarding work reorganization. The co-operatists are willing to introduce team forms of work organization which reduce the number of job classifications in the typical local contract, and have production workers take on inspection, minor maintenance, repair and some material handling tasks. These team systems are often accompanied by pay-for-knowledge systems which increase pay as competence is proved in a wider variety of tasks. In these team systems the work group informally regulates some of the issues previously administered through contractual language such as task assignment, overtime and vacation allocation, and some complaint and disciplinary actions.[4] For example, at the Wheeling-Pittsburgh and LTV steel companies steps are underway to replace supervisors with team leaders who are members of the union.[5] A similar approach is outlined in the Saturn agreement where the work team unit is only the lowest of a series of committees which make up the organizational structure and where the union has representatives on each of these committees (Saturn Corporation, 1985). For the co-operatists, as discussed below, the use of teams is linked to the introduction of more flexible technology which they believe works best with decentralized decision-making and more direct input from production and skilled workers.[6]

The militants oppose team systems and related committee and pay structures. They argue that these procedures are basically mechanisms to speed up work and are used by management so as to communicate directly with the workforce with the purpose of weakening workers' solidarity and loyalty to the union. The militants see the replacement of contractual regulation and detailed classifications as a step toward a more refined form of Taylorism and as a part of a movement to company unionism (Parker, 1985). For example, in his opening speech to the 1985 UAW-Canada convention, President Robert White warned that the Saturn agreement was a dangerous step toward enterprise (and weak) unions (White, 1985: 38–9).

The viewpoint of the two camps toward work reorganization is closely related to their views on how workers and unions should participate in decision-making. The co-operatists see team systems as the shopfloor component of enhanced worker and union participation in business decisions. The co-operatists also favour putting union representatives on corporate boards and other joint committees that have a say in either shopfloor or corporate planning.[7] Lynn Williams, for example, states, 'so far we've considered worker participation primarily on the lowest or shopfloor level. We need to be more daring. Bargaining unit employees have talents and ideas to contribute at middle-management levels as well. . . . Why shouldn't the Union be consulted on trade, investment, tax or environmental problems'? (Williams, 1986).[8]

Don Ephlin provides another illustration of this approach. He agreed to mutual growth forums in the 1982 UAW-Ford contract and the elaborate committee structure proposed for Saturn. Ephlin also has favoured the kind of plant level committee structure that exists at GM's Fiero plant in which the chairman of the local union's bargaining committee (the head of the local union in the Fiero plant) participates as a member of the plant manager's staff, attends all staff meetings, and has access to all the information the plant manager sees (Kochan, Katz and McKersie, 1986: 161–2).

Consistent with their views on pay concessions and work reorganization, the militants view the new participatory processes as shams that are part of a managerial strategy to co-opt and pacify unions. In their view, quality of working life and related programmes are essentially vehicles to speed up work either directly by getting workers 'voluntarily' to increase work effort or more subtly to divide and manipulate the workforce. To the militants, the proper forum for labour–management interaction is collective bargaining. In the words of William Winpisinger, 'programs not based on the collective bargaining relationship undermine the basic element of true democratic participation in the determination of working conditions. . . . Where there are real problems we will work with management through the already existing structure of inplant union representatives. . . . Why do we need some new organization when one already exists to handle these matters of mutual concern?' (Winpisinger, 1984: 200).

The militants thereby end up as staunch defenders of the traditional form of work organization and formal contractual regulation. For example, they oppose fewer job classifications on grounds that this will open the door to supervisor favouritism and speed-up (Parker, 1985). Similarly, the militants oppose having work groups resolve disputes more informally, or self-manage work on the grounds that this will create inter-worker rivalry and open the way to greater managerial control. When given the choice between experiments that involve work reorganization and the traditional hierarchical system, the militants prefer the traditional system.

One might expect that these different views of pay concessions, new forms of work organization, and industrial relations would lead the co-operatists and militants to support radically different government policies toward industrial adjustment. In fact, the political platforms of the two camps are very similar. Most of the co-operatists and militants favour local content legislation and more aggressive trade policies with the Japanese,[9] together with federal and state governmental 'industrial policies' of one sort or another.[10] Both also are in favour of more governmental income and training assistance for workers displaced by industrial change. Furthermore, in the US, both the co-operatists and militants generally look to the Democratic Party to accomplish these changes.

One area where the two camps disagree about government's role concerns training policies. At Ford and GM, Ephlin negotiated company-funded new training programmes, funded by corporate contributions, focused on the retraining of workers displaced by new technology and tuition assistance to help displaced workers find jobs either inside or outside the auto industry. In contrast, Robert White eschews these programmes on grounds that training is a governmental and not a private responsibility.

At the heart of this difference in approach to training is a more profound difference of opinion between the co-operatists and militants. The militants seem fearful that work reorganization and other steps taken by labour and management to improve competitiveness will ultimately lead the government to abdicate its responsibility to use macroeconomic, trade and social regulation to sustain suitable working conditions. The co-operatists would respond to this charge by claiming that the reforms they recommend for industrial relations must be supplemented by governmental action and are not a substitute for supporting government policies.[11]

There are some militant unionists who favour a more direct governmental response to economic pressures. Some activists typically at community levels, favour community ownership (a local variant of nationalization). For instance, a coalition of union and community groups, the Tri-State Conference on Steel, has waged a campaign to establish a governmental authority that would facilitate the transfer of ownership and continued operation of steel plants in the Pittsburgh area (Bureau of National Affairs, 1985a).

The Assumptions Underlying the Co-operatist and Militant Positions

In part the co-operatists and militants favour different strategies because they hold different assessments of the objective economic and technological conditions that labour confronts. Yet the major source of disagreement is the different views (ideologies) each side holds regarding the possibilities for labour–management interaction.

One objective condition the co-operatists and militants often disagree about is the nature of product demand. The co-operatists accept the negotiation of pay concessions because they believe the demand for unionized labour has declined and has become more price elastic. In the face of these shifts it makes sense to lower pay as a way to expand employment. The militants, on the other hand, are not convinced that the demand for union labour is very price elastic. In their view the oligopolistic structure of industry allows firms to increase or divert profits if afforded declines in labour costs, a causal chain which severs any link between labour costs and employment. The militants would concede that pay cuts lead to short-run employment gains in certain cases and yet still oppose those pay cuts on grounds that concessions by any firm or industry will spread to other industries through managerial whipsaw tactics and in the end lead to a broad lowering of labour standards. Militants would also question any macroeconomic association between lower labour costs and employment (most often on Keynesian grounds) and argue that any employment gains are not worth the price of the alleged long-run decline in labour standards.

It is interesting to note that some of the difference in views regarding product and labour demand springs from differences in economic conditions across industries. Some of the militant trade union leaders are located in industries where employment has not declined substantially and where the elasticity of demand for labour has not increased significantly. For example, in the airline industry the pilots' union, the Airline Pilots Association, has been much more willing to agree to pay concessions than have the maintenance engineers, represented by the International Association of Machinists (IAM). Cappelli (1985) argues that this difference in policy springs in part from the fact that the pilots have a lot to lose if they are laid off because their external market opportunities are less attractive, and they face steep internal salary structures. The ground and maintenance airline workers represented by the IAM, on the other hand, face attractive job opportunities in the labour market and do not receive large seniority salary premiums that would be forfeited if they were forced to change firms. So the fact that Winpisinger, president of the IAM, has been more opposed to pay concessions in airline bargaining than the leaders of the pilots' (and flight attendants') unions, may in part arise as a consequence of these different market conditions.

A similar case holds for the Canadian versus US branches of the UAW. The Canadian UAW has been more resistant to pay concessions and to experimentation with work reorganization and participation processes as compared to the US UAW. Again, market conditions support this policy difference. The Canadians faced a much smaller decline in vehicle sales and employment over the 1980–3 period as compared to the Americans. Furthermore, the prospects for the bargaining strength of the Canadian UAW could be construed as more advantageous than those facing the US branch of the UAW. The US UAW appears to be in a relatively weaker bargaining position because it faces a substantial and growing non-union sector in the auto parts industry and the opening of non-union assembly plants by Japanese parent firms.[12]

The militants and co-operatists also disagree regarding the objective nature of current technological changes. Many co-operatists favour work reorganization and increased worker and union participation because they believe these processes are closely linked to the introduction of flexible production techniques. The co-operatists maintain that the expanded use of microelectronics and heightened instability and uncertainty in market demand have created an opportunity for non-Taylorist forms of work organization. Following this argument, the co-operatists believe it is the presence of this technological and market push for more decentralized production techniques that sets the recent experimentation with new forms of work organization and worker participation apart from earlier reforms along related lines.[13]

There is substantial variation, however, in the degree to which co-operatists accept the argument put forward by Piore and Sabel (1984) that a shift toward flexible specialized manufacturing is underway. Some co-operatists would claim that the movement toward teamwork is largely in response to Japanese competitive pressures, and that manufacturing technology is still basically mass production (assembly-line) oriented. Nonetheless, even these co-operatists would see a market motivation for team and participatory structures.

The militants have a very different interpretation of the technological trajectory. In their view microelectronic technology is being used to deskill the workforce further and provide management with new surveillance and control powers. The militant interpretation is that new technology is being implemented in the same way that Braverman (1974) argues technology was used in the 1950s and 1960s, and Montgomery (1979) at the turn of the twentieth century (Shaiken, 1984; Noble, 1977). The militants might concede that it is possible for microelectronic technology to be used in a non-Taylorist fashion, and that technology used in this way could be more efficient than Taylorist production methods, but they believe this is not occurring (and probably cannot prevail under capitalism). So whereas the co-operatists see something very new in microelectronic technology, the militants see something very familiar.

There are also militants who do not agree with either the Braverman or Piore and Sabel interpretations of the consequences of technological change. These individuals are attracted to the traditional system of work organization and labour relations because they believe it provides a satisfactory (and inevitable) compromise between management's and labour's needs. These 'militants' are particularly sceptical of the co-operatists' view that work reorganization involving more direct input by workers into production decision-making is feasible in the face of what the militants believe is an overpowering inevitable clash of interests between labour and management.[14]

The most substantial difference between the co-operatists and the militants concerns the ideological premises held by each side. The co-operatists believe that it is possible for workers and unions to gain more meaningful participation in business decisions even under a capitalist system of property relations. To the co-operatists, although management may not in principle like to share more power with workers and unions, management is currently led to work restructuring by economic and technological pressures in certain industries.

The co-operatists differ in their predictions and preferences regarding how far such power-sharing could proceed. To some, the experimentation currently underway with team work systems and greater union participation in strategic business decisions might eventually join with other political changes so as to produce broader social change. Such a possibility is alluded to in Piore and Sabel's discussion of how flexible specialization could develop into a system of 'yeoman democracy'. Other co-operatists are less optimistic. In their view work reorganization is an important step toward a more complete form of industrial democracy, but current changes are just another incremental step along a road that unionism has been pushing for a long time (Bluestone, 1986).

The militants do not believe that a meaningful expansion of worker and union participation in decision-making is currently feasible. They point to the fact that experimentation with work reorganization and alleged participation processes are occurring at a time when the non-union sector is growing rapidly in the US and when at the national political level, a conservative movement is eroding many of the key features of the post-war New Deal system of social regulation in the US.

Yet within the ranks of the militants are two very different political visions regarding what sort of labour relations would be desirable. Some of the militants (the left wing) are in favour of the sort of radical social change Piore and Sabel discuss, but do not think it is feasible unless broader political changes were first to occur. There are also politically conservative militants, one could call them the 'preservation-ists', who favour traditional industrial relations and work organization because they believe these systems yield satisfactory outcomes to both labour and management.[15] The preservationists would oppose broader political change even if such change were feasible. In collective bargaining, preservationists will accept pay concessions and work rule changes so as to expand employment. In this way the practical policies endorsed by the preservationists are closer to the co-operatists' position than they are to the militants' policies.

Where individuals on the political right and left are joined in the militant camp is through their opposition to experiments that reorganize work and labour relations.[16] Both groups end up favouring the *status quo*. The militants on the left do so because they do not believe anything else is feasible. The militants on the right do so because they like the traditional system.[17]

Compromises in Practice

The co-operatists sometimes militantly oppose management's demands. Meanwhile, the militants sometimes accept pay concessions and work restructuring. Each side now faces the challenge of defining what and where compromise is appropriate. The definition of appropriate compromise puts pressure on each side to articulate their respective vision of the future. Yet to date both the co-operatists and militants in the North American labour movement have failed to articulate such a vision, thereby exacerbating the problems each side confronts as they face new economic pressures and management's strategic actions.

For the co-operatists the critical issue is how to avoid a situation in which their willingness to experiment with new work and industrial relations systems becomes wholesale acceptance of management's plans and a wholesale abandonment of any national labour standards. Local unions face this problem when, hard pressed by management's threats of outsourcing, they have to decide which pay and work rule changes they will accept. An answer could come in the form of local application of a national union plan for work reorganization and enhanced worker and union participation. But no national union or national union leader has articulated such a plan, even in industries where extensive local experimentation is already underway. Lacking such a plan, national unions have on occasion blocked locally initiated changes when it appears that these local actions either lack sufficient safeguards or greatly exceed changes made elsewhere. In this manner imaginative local experiments sometimes are ruled out only because their implications are misunderstood.

Another problem created for the co-operatists by the lack of a clear plan is the difficulty they face in answering militants' charges that local experiments merely amount to co-optation and a violation of national standards. Again, without a national plan that articulates a purpose and links local efforts, co-operatist union leaders are hard pressed to distinguish acceptable and unacceptable compromise.

The militants face symmetric problems. For them problems arise when, under the threat of further employment loss, the militants decide that certain pay concessions or work rule changes are acceptable. The IAM, for example, has accepted pay cuts and work rule concessions in some airlines and aerospace agreements. In the auto industry, Robert White and Stephen Yokich, the latter as head of the Ford Department of the UAW, each have accepted concessions in certain situations. Each, for example, has expressed a willingness to agree to work rule concessions in new plants to be built by Japanese car companies as a way to ensure that those plants are built in the US or Canada.[18] Both White and Yokich, however, oppose the sort of work reorganization and participatory programmes favoured by Ephlin and other co-operatists in the auto industry. Presumably, where they grant concessions the militants are hopeful that the concessions would prove to be only temporary, or at least these limited changes avoided the alleged co-optive elements of the co-operative approach.

The militants, like the co-operatists, face the difficulty of defining when such compromise is acceptable. Again, the answer could lie in an articulated national plan. But such a plan has not been outlined. In the absence of a clear plan the militants face the danger that their position can be characterized as a lack of imagination and foresight. When opposing the militants, the co-operatists can appeal to the workforce on the grounds that a co-operative approach produces more employment and claim that the militants fail to confront unattractive, but at the same time, real economic pressures.

The militants also face the danger that their position can be characterized as a self-fulfilling prophecy. Since they oppose current experiments in work reorganization, the militants emerge as defenders of traditional Taylorist forms of work organization. And since the militants rule out the possibility of enhanced worker and union input into managerial decision-making, they appear to be defenders of a narrow form of business unionism. As discussed above, the political conservatives among the militants would find nothing unsettling about this characterization. But the left wing of the militants certainly finds itself in an uneasy position as the defender of Taylorism and business unionism. To avoid this challenge the militants, like the co-operatists, need a plan for the future.

Notes

I have benefited greatly from discussions with Jeff Keefe, Mike Parker and Chuck Sabel.

1. The terms co-operatists and militants are not completely satisfying. As discussed below, militants co-operate in some workplace changes and co-operatists favour work reorganization and not necessarily co-operation with management.
2. For detailed descriptions of these contracts, see Bureau of National Affairs (1985b, 1986). Williams also supported 1983 steel pay concessions when he was secretary-treasurer of the USW. Ephlin negotiated the 1982 agreement with Ford, the 1984 agreement with GM and the Saturn agreement with GM. These agreements all included pay concessions, although unlike the steel and many other recent union agreements, the auto agreements merely slowed the rise in compensation and did not include pay cuts (Katz, 1985). The co-operatists would not refer to these agreements as concessions, since from their viewpoint the agreements include many gains such as explicit new job security programmes, increased worker and union participation etc. Union gains in concession bargaining are described in Cappelli (1984).
3. For example, Victor Reuther, brother of the late Walter Reuther, warned that by linking worker compensation to economic performance the Saturn agreement was bringing a return to the piece rate system (Serrin, 1985).
4. Team systems in the auto industry are described in Katz (1985) and Katz and Sabel (1985).
5. The 1985 contract between the Wheeling-Pittsburgh Steel Corporation and the USW put union officials on the corporate board of directors and created joint production boards and a 'Joint Strategic Decisions Board' (Serrin, 1986).
6. For a discussion of the links between team systems and flexible manufacturing technologies see Piore and Sabel (1984), Katz and Sabel (1985).
7. A strong endorsement of broad forms of worker and union participation in decision-making is provided by unionists in the communications industry in Straw and Heckscher (1984).
8. For the views of another trade union leader who favours broad union participation in corporate decision-making see Joyce (1984).
9. Some co-operatists oppose local content and other protectionist policies on the grounds that they do not provide long-run solutions to heightened international competition and have harmful side effects.
10. For example, Lynn Williams states, 'government must play a vital and active role in the restructuring and modernization of our industrial base' (Williams, 1986).
11. Again, Lynn Williams states, 'the co-ordination of public and private efforts is required if we are to deal effectively with the new realities that challenge us in this global economy' (Williams, 1986).
12. Lower Canadian labour costs are a product of the depreciation of the Canadian currency relative to the US dollar and the presence of national

health insurance in Canada which lowers private health care costs to the auto firms. [Editors' note: the tensions noted in this discussion have now resulted in the breaking away of the Canadian section of the UAW to form the Canadian Auto Workers.]

13. The link between technological change and participatory processes is stressed in Katz and Sabel (1985) and Hirschhorn (1984).

14. It is something of a misnomer to call this position a militant strategy. The philosophic differences within the militant camp are discussed in more detail below.

15. A clear statement of this position and scepticism regarding worker and union participation programmes is provided in Dunlop (1986).

16. It is difficult to find clear expressions where the left and the right militant factions join together in opposition to work reorganization and labour relations change. My observations here draw from training sessions I participated in where local union leaders discussed the course of industrial adjustment.

17. The possibility of such alliances is discussed in Katz and Sabel (1985).

18. For example, in 1986 Robert White signed an agreement with the Suzuki Motor Company providing for the use of teamwork units and other flexible work rules at Suzuki's new car assembly plant (Slotnick, 1986).

References

Bluestone, I. 1986. 'Changes in US Labor–Management Relations'. *Proceedings of the Thirty-eighth Annual Meeting of the Industrial Relations Research Association*. Madison, Wisconsin: Industrial Relations Research Association, 165–70.

Braverman, H. 1974. *Labor and Monopoly Capital*. New York: Monthly Review Press.

Bureau of National Affairs. 1985a. 'Local Governments and Unions in Pittsburgh Area Mobilizing to Prevent Further Loss of Steel Jobs'. *Daily Labor Reports*. no. 20, January, 30, A–2,3.

Bureau of National Affairs. 1985b. 'Tentative Accord with Steelworkers Could End Three-Month Strike at Wheeling-Pittsburgh'. *Daily Labor Reports*. no. 200. 16 October A–15,16.

Bureau of National Affairs. 1986. 'Steelworkers Accept Pay Deferral at LTV Steel in Exchange for Gains in SUB, Other Provisions'. *Daily Labor Reports*. no. 23. 4 February, A,9.

Cappelli, P. 1984. 'Union Gains Under Concession Bargaining'. *Proceedings of the Thirty-Sixth Annual Meeting of the Industrial Relations Research Association*. Madison, Wisconsin: IRRA, 297–305.

Cappelli, P. 1985. 'Competitive Pressures and Labor Relations in the Airline Industry'. *Industrial Relations*. Vol. 24, no. 3, Fall, 316–38.

Dunlop, J. T. 1986. 'A Decade of National Experience'. *Teamwork: Joint Labor–Management Programs in America*. Ed. J. Rosow. New York: Pergamon Press, 12–25.

Hirschhorn, L. T. 1984. *Beyond Mechanization*. Cambridge, Mass.: MIT Press.

Joyce, J. T. 1984. 'Codetermination, Collective Bargaining, and Worker

Participation in the Construction Industry'. *Challenges and Choices Facing American Labor*. Ed. T. Kochan. Cambridge, Mass.: MIT Press, 257–70.

Katz, H. C. 1985. *Shifting Gears: Changing Labor Relations in the US Automobile Industry*. Cambridge, Mass.: MIT Press.

Katz, H. C., and F. Sabel. 1985. 'Industrial Relations and Industrial Adjustment in the Car Industry'. *Industrial Relations*. Vol. 24, no. 3, Fall, 295–315.

Kochan, T. A., H. C. Katz, and R. B. McKersie. 1986. *The Transformation of American Industrial Relations*. New York: Basic Books.

Montgomery, D. 1979. *Workers' Control in America*. Cambridge: Cambridge University Press.

Noble, D. 1977. *America By Design*. New York: Knopf.

Parker, M. 1985. *Inside the Circle: A Union Guide to QWL*. Boston: South End Press.

Piore, M. J. and C. F. Sabel. 1984. *The Second Industrial Divide*. New York: Basic Books.

Saturn Corporation. 1985. 'Memorandum of Agreement Between Saturn Corporation and UAW'. July.

Serrin, W. 1985. 'Saturn Pact Assailed by a UAW Founder'. *New York Times*. 28 October, 18.

Serrin, W. 1986. 'Early Signs of Promise in Union "Partnership" at Steel Company'. *New York Times*. 7 April, 8.

Shaiken, H. 1984. *Work Transformed*. New York: Holt, Rinehart and Winston.

Slaughter, J. 1983. *Concessions and How to Beat Them*. Boston: South End Press.

Slotnick, L. 1986. 'Will Canadian workers accept Japanese ways'? *The Globe and Mail*. 6 September, D–1.

Straw, R. J., and C. C. Heckscher. 1984. 'QWL: New Working Relationships in the Communication Industry'. *Labor Studies Journal*. Winter, 261–74.

White, R. 1985. Speech to the UAW-Canada National Convention. 4 September, unpublished transcript in possession of the author.

Williams, L. R. 1986. 'Collective Bargaining Prospects'. *Proceedings of the Thirty-Eighth Annual Meeting of the Industrial Relations Research Association*. Ed. B. Dennis. Madison, Wisconsin: IRRA, 14–25.

Winpisinger, W. W. 1984. 'Comments'. *Worker Participation and American Unions: Threat or Opportunity?* Ed. T. A. Kochan, H.C. Katz and N. Mower. Kalamazoo: W. E. Upjohn Institute for Employment Research, 199–202.

14

The Australian Metalworkers' Union and Industrial Change: A Labour Movement Offensive

Stephen Frenkel

The Amalgamated Metalworkers' Union (AMWU) is the largest manual union in Australia. With a predominantly male craft worker membership and a left-wing leadership, the AMWU plays a prominent role in developing strategy for the wider union movement. It does this directly, through the influence of its leading officials within the peak union body, the Australian Council of Trade Unions (ACTU), and indirectly, through the pattern-setting nature of its awards and agreements. The union's experience of the trade cycle and the current recession has tended to be sharper than that of most of its counterparts, a factor which las led AMWU officials to question traditional approaches to labour betterment and to concentrate on devising new strategies and tactics.

My intention in this chapter is to show how industrial democracy is becoming an increasingly important element in the metalworkers' programme for industrial regeneration. To understand this, it is necessary to examine employer action in the context of the present Labor government's approach to economic policy-making. It is in response to management's cost-cutting and efficiency-oriented strategies under Labor that the AMWU has been paying greater attention to workplace industrial relations.

From Boom to Recession: Union Strategy in Transition

Between 1968 and 1975 the Australian economy was operating under conditions of relatively low unemployment and high inflation. The

AMWU was in the vanguard of a movement which pursued improvements in wages and conditions largely by decentralized bargaining backed by industrial action. This approach was formalized, in part, by reference to union policy on industrial democracy. This policy, devised and amended in the early 1970s, emphasized the need for strong shop steward organization and intervention in managerial prerogatives. It reflected the radical ideas popularized at the time while also serving as a safeguard against the allegedly excessive autonomy characteristic of the British shop steward movement. Despite considerable discussion of industrial democracy, union policy was only partly implemented: shop steward organizations did develop but they rarely sought to broaden systematically the bargaining agenda. This was largely on account of the inflationary circumstances which encouraged and facilitated continuous, unco-ordinated sectional action over wages and conditions. Moving from one strike to another, union organizers found it difficult to focus attention on control issues. Moreover, management vigorously defended their prerogatives, preferring to grant wage rises which could usually be passed on in the form of higher prices.

By 1976 circumstances had changed dramatically. A Conservative government, dedicated to reducing inflation by increasing unemployment and containing militant union action had assumed power (Frenkel and Coolican, 1984: 17–29). The economy began to move into recession. This had an adverse effect on the metal industry with five times as many skilled electrical and metal workers seeking employment than there were vacancies available. The plight of less skilled workers was nearly four times worse (Frenkel and Coolican, 1984: 93). In addition, earnings relativities in the metal industry, with few exceptions, began to fall behind those of workers in sectors less influenced by the downturn in the economy. The AMWU now faced a major challenge, given its weak position in the labour market and lack of power in the political arena. For a while it sought compromises with the engineering employers through industry-level bargaining. Owing to a short-lived improvement in the economy, associated with an unexpected boom in the minerals sector, the AMWU and other metal industry unions won major gains in 1981, including a reduction in weekly working hours from 40 to 38. As a *quid pro quo*, the unions agreed to abide by a 'no extra claims' clause which, in effect, prohibited further pay and conditions claims (and hence industrial action over these matters) for two years. This set a precedent which was followed by other Australian unions who, like the AMWU, had little labour market power in any case.

Privately, AMWU leaders acknowledged that under depressed conditions wage increases might lead to higher unemployment. By the end of 1982 the unemployment rate had climbed to 9.2 per cent compared to 6.0 per cent two years earlier. There was now an urgent need to replace collective bargaining with a new strategy. For the traditional alternative – centralized, nation-wide arbitration – offered very little to the unions, because of restraints observed by the main industrial tribunal, the Australian Conciliation and Arbitration Commission (ACAC) under pressure from employers and the federal government.[1] Metalworkers' leaders, supported by officials from other unions, came to the view that the only solution was political: to struggle for the election of a Labor government committed to fighting inflation and unemployment simultaneously. The AMWU developed a policy which, with some later modification, was endorsed by the ACTU and the Australian Labor Party as a basis for electioneering and subsequent action should the Labor Party succeed in the 1983 elections.

The Accord, as this policy has come to be known, involved the maintenance of real wages over time through nation-wide wage indexation and improvements to the social wage (health, education, superannuation, welfare services, etc.) by fiscal means. In exchange, the unions gave assurances to forego wage claims outside wage indexation. The unions also supported the restructuring of industry on condition that they played a meaningful role in the planning process which would precede major industrial change. It was on the basis of the Accord that, in March 1983, the new Labor government under Prime Minister Hawke began a new era in industrial relations.

Employee Reaction in the Context of Labor Party Strategy and Economic Recession

Although wage *increases* were tightly controlled, initially by means of an agreed six-month wage pause, and later, from September 1983, by national wage indexation, employers faced an unprecendented situation. Increasingly competitive markets made it difficult to improve profit-ability, while tribunal and statutory control over labour costs made it impossible in the short term to reduce these costs and win orders on the basis of lower prices. Lack of management control over labour costs was and continues to be not just a question of the wage component: other costs such as redundancy pay, holiday pay, long service leave pay, workers' compensation for injury – all these so-called labour on-costs – are regulated by law. In addition, employers face claims for superannuation (discussed later in the chapter).

How then have employers responded to this profit squeeze? At the collective level – through their employer associations – they have tried to restrain ACAC from awarding wage increases. Together with other employer organizations, they opposed the ACTU claim for redundancy pay and mandatory consultation over major technological and organization change. They have also sought federal and state government support for tax reductions.

For the most part the employers have been unsuccessful: for the time being the state remains sympathetic to labour in both federal and state government spheres. Employers have thus been encouraged to concentrate on improving profitability through the rationalization of production units. This will be discussed under two headings: company restructuring and changes at the point of production.

Company Restructuring
Restructuring takes many forms with the three most common types being the introduction of new technology or new operational methods; broadening of a firm's product range; and substitution of locally-produced components or finished products by imports. According to an employer association survey (MTIA/Commonwealth Bank, 1985), 18.2 per cent of metal industry capital goods firms introduced new operational methods or new technology in 1984, increasing to 27.3 per cent in 1985. In 1984, 10.5 per cent of firms surveyed reported an increase in their product range; this proportion more than doubled (22.4 per cent) in 1985. Imported components and finished goods also increased, with 13.6 per cent of capital goods firms registering an increase in 1984 and 16.9 per cent reporting the same in 1985. The picture for metal industry consumer goods firms is similar: 18.5 per cent reported new technology or new operational methods in 1984, 29.3 per cent in 1985. In terms of increasing the range of products, 12.2 per cent recorded an increase in 1984, 22.4 per cent in 1985. The increase in imported components or finished products was much less substantial: from 15.0 per cent of metal industry consumer goods firms to 18.4 per cent. Of the nine separate forms of restructuring referred to in the survey,[2] over a third of metal industry firms (capital and consumer goods producers) were engaged in one or more forms of restructuring increasing to over a half in 1985, with close on three-quarters anticipating some kind of structural change in 1986.

In addition to the forms of restructuring referred to above, it is noteworthy that mergers and the relatively rapid demise of small firms have resulted in a growing concentration of corporate control in the metal industry, with 37.7 per cent of value added in 1983 (most recent data available) being produced by the largest eight enterprise groups (ABS, Cat. No. 8307.0). As control becomes more concentrated, average plant size is declining. Thus in 1978–9 there were 49 persons employed in the average metal industry plant; by 1984–5 this figure had declined to 39 (ABS, Cat. No. 8202.0).

Changes at the Point of Production: (1) Economizing on Manufacturing Costs

The drift towards smaller plants has been influenced by management's search for greater productivity through 'natural wastage' and redundancies, particularly in larger plants. Employment fell by 17 per cent in the year to May 1983, with a further 5.2 per cent reduction in the year ending February 1984 (MTIA, 1984). Although the rate of job loss has since moderated, employment is unlikely to return to its pre-recession level.

Retrenchments over the past few years have not affected the metal industry workforce in a uniform way. Worst hit have been semiskilled workers whose numbers in the state of New South Wales declined by 38 per cent in the three-year period ending May 1984. Metal and vehicle trades people also fared badly with a reduction of 28 per cent over the same period (DEIR, 1985). Despite the recent improvement in the economy, including the metal industry, employers have been reluctant to engage additional full-time employees, preferring to increase overtime working and, where necessary, hire workers on the basis of fixed-term contracts (ABS, Cat. No. 6330.0; Kirby, 1985). Shortages of particular categories of workers have resulted in 'grade drift' (upgrading of jobs bringing more pay to incumbents) and job reclassifications in order to retain or attract scarce labour. However, the extent of wage drift has been very low, estimated at less than 1 per cent in 1985 (ABS, Cat. Nos 6301.0 and 6302.0).

Another possible way of reducing unit costs is by introducing new technology. However, most innovation has been limited in scope – embracing a small proportion of machinery and equipment in any particular plant – and limited in technological sophistication (ASTEC, 1985; Frenkel, 1985). It therefore makes sense to consider technological change and new working practices together, since the former is often associated with the latter.[3] But first it is necessary to ask why management have been reluctant to invest significantly in the more recent generations of microprocessor technology. There appear to be three main reasons: (1) lack of investment funds at low rates of interest; (2) uncertainty over future market trends; and (3) unfamiliarity with sophisticated computerized systems such as CAD/CAM and Computer Integrated Manufacturing. Nevertheless, as mentioned earlier in relation to restructuring, there has been a widespread tendency to tighten up on inefficient use of machinery and equipment and wasteful work practices. In particular, economies have been sought in inventory levels; machinery set-up and running times; component queuing and transit times for raw materials and semi-processed materials; quality control; and machine maintenance. Although many companies have approached these issues on an *ad hoc* basis, a growing number of firms, with encouragement from federal and state governments, are adopting a more comprehensive and systematic approach to raising productivity. This is known as 'just-in-time' with 'total quality control' (JIT for short).

JIT configurations are designed to produce and convey material, whether in a raw, semi-processed or finished form, precisely when required (Schonberger, 1982). Together with a greater emphasis on individual employee responsibility for quality, this system of production is capable of achieving significant unit cost reductions, particularly in relation to inventory and reject costs.

The introduction of JIT is usually accompanied by major organizational changes including alterations in plant layout. Supervisors are encouraged to play a more facilitating role with greater stress on employee motivation. At the same time, machine operators are required to assume greater responsibility for groups of machines, their role being enlarged to encompass simple preventative and remedial maintenance tasks and the application of diagnostic skills to maintain high quality standards.

More generally, JIT brings management closer to the shopfloor for it requires constant attention to production details. Management become more dependent on workers' active involvement, as the success of the firm is predicated on its ability to meet changing market demand through flexible product mixing at relatively low cost. Management therefore seek employee co-operation by attempting to widen workers' concerns beyond their immediate tasks, using devices such as informal productivity group meetings and quality circles (see below). Although this type of consultation is likely to gain the approval of employees who feel undervalued, it puts additional pressure on workers regardless of whether or not such schemes are voluntary.[4] This issue also raises the question of compensation for productivity improvements, something which management are reluctant to endorse, particularly when existing national wage principles militate against gain-sharing of this kind.

The attempt to harness employees' previously untapped commitment and expertise has been reinforced by several additional factors: job insecurity has encouraged worker receptiveness to 'help yourself by helping the firm' propaganda; publicity given to the Japanese model of labour relations, including fear of increased competition from Japan, has influenced workers, unions and management as has the ideological current of consensus decision-making publicized so frequently by the Hawke government. More concretely, the government has fostered worker participation schemes through the provision of 'seed money' for experimentation by firms, employer associations and unions. Indeed, the AMWU has been a major participant in this process.

**Changes at the Point of Production: (2) The New Industrial
Relations and Employee Involvement**
Several trends in workplace industrial relations are particularly evident
in large multi-plant companies. Management rationalization of organizational structures involves the following tendencies: (1) integration of
industrial relations and personnel sections or departments; (2) elevation
of this human resource function, with the incorporation of issues such
as employee career planning, motivation and regulation into strategic
plans; (3) a greater role for lower line management in dispute
prevention and resolution; guided by (4) codified rules and procedures
established by management, sometimes in conjunction with unions and
tribunals but more often in response to emerging trends in labour law;
finally, (5) the introduction of employee involvement schemes which
was touched on earlier. Before examining this aspect more closely, it
is important to note that there has been no explicit management
offensive against unions at workplace level. Manual union density
remains high, on average over 80 per cent for metal plants with 50 or
more employees (Frenkel, 1985). On the other hand, shop steward
organization and influence have been severely limited by the centralization of wage determination and constraints placed on workplace
bargaining by tight ACAC wage guidelines. Fears of redundancy by
workers, in conjunction with the tendency for employers to seek co-
operation directly from employees rather than through the medium of
union representatives, have also been important. For example,
according to a recent employer association survey of 846 metal industry
plants, slightly over half (51.7 per cent) of managers – covering nearly
two-thirds of employees in respondent plants – use management–
employee meetings to inform employees about company developments.
By contrast, similar meetings between management and union representatives were reported in only 23.1 per cent of plants, although these
accounted for slightly more than half (51.9 per cent) of employees.
Even when we confine our analysis to large plants of 100 or more
employees (which are also the most highly unionized), management–
employee meetings are apparently more frequently used than communications with union representatives.[5]

In the same survey, 29.7 per cent of metal industry workplaces
(employing 57 per cent of the total workforce covered) operated one
or more participation schemes in November 1984. The figures indicate
that such schemes are most common in larger workplaces. *Direct*
employee participation existed in 44.4 per cent of cases, particularly
in the form of productivity groups and quality circles. Regular
management–employee meetings are also frequently used. Of the
representative structures, safety committees (20.8 per cent) and joint
consultative committees (15.1 per cent) are the most popular.

It appears that direct participative forms have been growing most rapidly, with 51.4 per cent introduced in the three years immediately prior to the survey, compared to the equivalent figure of 35.1 per cent for representative schemes. Productivity groups and quality circles again feature prominently. The survival rate of these schemes is high: 89.9 per cent for direct participation, 95 per cent for representative structures. This suggests that management are satisfied with the operation of these schemes, a conclusion supported by additional data reported elsewhere (Frenkel, 1985).

A further point to note about developments in worker participation is that although these cover a minority of plants, albeit a majority of larger plants, management are by and large favourably disposed to extending certain forms of participation – particularly productivity improvement groups, management–employee meetings and safety committees – to plants where no schemes currently exist. On the other hand, the survey data indicate that management are highly ambivalent about *negotiating*, in contrast to *consulting* with worker representatives about changes in remuneration, technology and working practices. However, it is precisely over issues such as health and safety, redundancies, superannuation, work organization and work standards that employees wish to exert more control (AMWU, 1985).

With pressure to reduce unit costs and raise productivity, it is not surprising that the impetus for 'employee involvement' comes from management. This raises the question of union attitudes and responses to these developments. The other metal unions have been content to leave this issue in the hands of the AMWU. Its approach is shaped by values pursued by union leaders in conjunction with the union's increasing experience in this domain. Before examining this in greater depth, it is worth noting that the AMWU has been the most influential affiliate in shaping the ACTU's industrial democracy policy and devotes more resources to this area than any other Australian union. As we shall observe shortly, this not only implies something about the AMWU, it also indicates the relative paucity of commitment and involvement by other unions in key issues at workplace and enterprise level.

Industrial Democracy as an Element in the Metalworkers' Industry Regeneration Strategy

Among the most important issues facing the AMWU is the prospect of further redundancies in the metal industry. Over the longer term, there is the spectre of a substantially reduced manufacturing base, in part generated by technological change. In addition, the tendency towards smaller plants and more sophisticated management strategies provides employers with an opportunity to eliminate trade union influence at the point of production. The AMWU's response has been to seek union rights in decision-making over issues directly linked to job control as part of a wider strategy of industry regeneration. According to this view, such potentially *distributive* issues as job security, job design, health and safety, reskilling and training can only be satisfactorily resolved in the context of an effective *production* policy capable of ensuring that the metal industry improves its share of domestic and export markets. This approach presupposes union participation and influence in macroeconomic policy, including the operation of protective labour market programmes. Within the framework of the current accord between unions and government in Australia, the metalworkers seek to extend their influence to major production decisions through the mechanism of industry planning while pursuing the implementation of such plans, including employee protection, by means of industrial democracy. However practice falls considerably short of policy. In 1984–5 the AMWU, together with the other metal unions, attempted to develop regional industrial planning mechanisms for the highly industrialized western areas of Sydney and Melbourne respectively. These initiatives were subsidized by government grants which enabled regional shop steward committees to arrange seminars and conduct research with the assistance of external consultants of their choice. Two reports on trends in employment and output were compiled, one for each region. These included discussion of barriers to industrial regeneration and proposals for fostering technological change and improving economic relations between firms in each area. Managers were invited to co-operate in the research and associated discussions, but the response was desultory. Ultimately, nothing came of these attempts at union involvement in industrial planning, in part because of lack of active support by government departments, but more importantly because employers oppose the idea of planning, particularly where this process includes trade union participation.[6]

A second form of intervention attempted by the AMWU has been through the medium of federal government industry assistance plans. These are designed to provide financial assistance for restructuring specific industrial sectors such as steel, cars, shipbuilding and heavy engineering. Assistance is conditional on firms meeting production and price targets set out in the plans. The AMWU, in association with other unions, has sought to include additional assurances in these plans. Thus, in the shipbuilding plan currently under discussion, the unions demand access to financial information and involvement in decisions on technological change, job design, upgrading of skills, alterations in work practices, redeployment of labour and worker retraining. In return, they have indicated their willingness to discuss improvements in productivity and the introduction of 'broadbanding' (i.e. replacing the present complex structure of occupations and associated pay levels with a small number of broad occupational categories). This is seen as a precursor to enlarging workers' repertoire of skills and permitting management more flexibility in the deployment of labour. It is too early to say how successful this form of intervention will be.

This conclusion is less true of a third tactic, local content agreements. These highlight the metalworkers' preoccupation with unemployment and the future of the industry, although they are currently restricted to major natural resource developments such as the coal to oil project in Victoria. Local content agreements involve stewards monitoring the purchase of equipment and materials and negotiating for the replacement of imports by locally produced goods. Sometimes, as in the case of the coal industry in the Newcastle area of New South Wales, joint working parties have been established to examine future purchases with a view to increasing local content where this is economically feasible. The presumption is that this type of approach will minimize the possibility of industrial action over this issue.

With growing government financial and organizational support for advanced technology like CAD/CAM and new manufacturing techniques such as JIT, the AMWU has framed policies to ensure that unions exert their own priorities in the decision-making process.[7] This fourth element in the union's industrial democracy strategy requires union participation in tripartite monitoring and project committees operating at state government and workplace levels respectively. The metalworkers insist that any decision to proceed with this type of far-reaching change must have union approval. Specific rights are sought on such issues as job security, paid study leave for stewards, protection for shop steward organization, increases in skill levels, improved job design and disclosure of information. Reports from New South Wales and Victoria indicate that after some initial problems, union involvement in these projects has become more acceptable to experimenting firms. This is in part due to the support given to the unions on this issue by federal and state Labor governments and the force of the 1984 ACAC Termination, Change and Redundancy decision which requires all managements who are party to federal awards to disclose information and consult with the relevant unions prior to introducing major technological and organizational changes in the workplace.

A fifth path towards industrial democracy is the negotiation of agreements embodying new forms of co-operation and control. The metalworkers have used the government's industrial democracy grant scheme not only to survey their members on this issue but to develop a kit for shop stewards and organizers. However, attempts to negotiate industrial democracy agreements with several specially selected multi-plant companies have not been successful (although such efforts are still in their infancy). This is because few companies have been willing to explore this issue with the union. At one firm the union did succeed in obtaining an agreement to establish consultative committees at two sites. Complementary provisions in the agreement include: (1) opportunity for stewards to meet in company time to prepare their proposals; (2) steward training on issues agreed by the committees; (3) access by stewards to consultants of their choice, again subject to a majority vote in the committee; (4) equal numbers of management and union representatives with a rotating chairperson from each side; (5) a wide agenda, with no issues expressly excluded; and (6) the functioning of the committees to be evaluated in the near future.

The sixth and final aspect of the AMWU's approach to industrial democracy is also in its infancy; the recent establishment of a Manufacturing Unions Superannuation Trust (MUST) which seeks to compete with employer-controlled superannuation schemes. MUST has been facilitated by the revised Accord guidelines negotiated in 1985 by the ACTU and the federal government. These provide *inter alia* for the introduction of industry-based schemes and the equivalent of a 3 per cent wage increase in the form of superannuation benefits.

The connection between union-controlled superannuation funds and industrial democracy lies in their relationship to industry development, a policy area which has assumed increasing importance, not only with the metalworkers but also with the wider union movement. This is because Australia's terms of trade have been deteriorating, implying that traditional exports of primary commodities cannot be relied upon to resolve the country's growing foreign debt and balance of payments problems. The unions and government agree that the manufacturing sector – the metal industry being the largest component – will need to play a far more prominent role in import substitution and export expansion.

This requirement is made all the more urgent by the lacklustre response of the economy to the 1985 devaluation of the Australian dollar. The 30 per cent reduction in Australian prices relative to the country's major trading partners has, as yet, failed to boost exports significantly while the burden of imports continues to increase. The balance of payments problem represents an immediate threat not only to the Accord strategy but also to the election prospects of the Labor government and hence the political power of the union movement.

The AMWU and Australian unions more generally realize the importance of adequate and carefully targeted investments in the manufacturing sector. Through control over superannuation trust funds, they see a means of fostering industry development which would go some way to redressing the balance of payments constraint and providing job security and better employment prospects for present and future union members. But this approach is vigorously opposed by employers both inside and outside the High Court. It is viewed as a serious challenge to the free enterprise system.

Conclusion

This chapter tried to show how closely union strategy is attuned to changes in the political economy. Thus in the boom period of the early 1970s the metalworkers concentrated their attention on the labour market through the use of collective bargaining backed by direct action. The structural conditions which facilitated an increase in union power paradoxically impeded the union's capacity to broaden the political agenda. In essence, the anarchy of the labour market not only encouraged union leaders to pursue their main objectives through the market, it also restricted the range of issues over which they struggled; in short, economism reigned supreme. From the mid-1970s however, we see a new scenario unfolding, characterized by growing unemployment and continuing (albeit reduced) inflation. This, together with a hostile Conservative government, induced the AMWU to search for a new economic and industrial relations strategy. In the meantime union leaders retreated from the tactics of bargaining and direct action towards wage determination through centralized arbitration. This quasi-political process came to be relied upon more strongly following the brief upturn in the economy in 1980–1, which in turn occasioned a flurry of bargaining activity leading to the pace-setting 1981 Metal Industry Award. Facing a weak labour market, although with some measure of protection from ACAC against real wage reductions, AMWU leaders concluded that progress could only be made by replacing the government with a sympathetic Labor administration.

Election of the Hawke government in early 1983 constituted something of a victory for the AMWU; for the Accord strategy on which the government's macroeconomic policy rested reflected the approach developed by the metalworkers a year earlier. It was the conjuncture of deep recession and a particular labour movement economic strategy that encouraged metal industry employers to rationalize their production units, streamline production and seek employee compliance. This was the context to which the AMWU responded by elaborating an industrial development strategy in which industrial democracy, including control of technological change, has assumed increasing importance. This has occurred precisely because such policies simultaneously express and seek to counteract union members' fears regarding job security and skill obsolescence and union leaders' concerns over the weakness of workplace unionism and the threat of further deindustrialization. It is the connection between policies like these at the point of production and those at industry and the macroeconomic level, based on a strong union–Labor Party foundation, which justifies the term 'labour movement offensive' in the title of this chapter. This is not to say that the AMWU will succeed in attaining its main objectives: industry policy continues to be a matter of contention between the unions and the government while Australia's weak balance of payments position has put pressure on the government to adopt restrictive economic policies. Nevertheless, an outline of a very different political economy is now apparent and on the political agenda. The challenge facing the Metalworkers and other Australian unions is now to preserve the present relationship between the unions and the Labor government, using it as a basis for meeting the immediate aspirations of the workforce, while simultaneously mobilizing rank-and-file support for the kinds of strategies outlined in this chapter.

Notes

1. The Commission is the most important of several industrial tribunals regulated by statute. It is empowered to conciliate and arbitrate issues on the request of one or more parties where a dispute is found to exist. Its jurisdiction is limited by the High Court's interpretation of what constitutes an 'industrial issue'.
2. These nine forms of restructuring are: introducing new technology or new operational methods; commence/increase importing of parts, components or finished products; increase the range of products manufactured locally; reduce the range of products manufactured locally; diversify into non-manufacturing activities; engage in mergers, takeovers or joint ventures; close down manufacturing establishments; open new manufacturing establishments; and set up off-shore production facilities.
3. In a survey of 315 metal industry plants employing 50 or more employees (Frenkel, 1985), respondents were asked about their plant's experience of major changes over the five-year period ending in mid-1984. In two-thirds of the 21.0 per cent of plants where major technological change (affecting 25 per cent or more of the workforce) had taken place, this was accompanied by some other major change to the firm's organization structure or product line. This would seem to imply that alterations in working practices are frequently accompanied by or are the result of technological change.
4. Quality circles are voluntary but failure to comply with management expectations is likely to lead to subtle forms of discrimination against non-participants.
5. Well over half (57.7 per cent) of the survey respondents in plants with 100 or more employees reported the use of management-employee meetings compared to 46.2 per cent who stated that they used management–union meetings to communicate with their workforce.
6. In 1984 the metal trade unions commissioned a detailed analysis and policy proposal from a respected academic research institute. The ensuing report entitled *Policy for Industry Development and More Jobs* supported the unions' strategy of intervention in capital and product markets. The document was opposed not only by employer groups but also by key government departments whose senior personnel are totally committed to neoclassical economics.

7. The metalworkers' general policy on technological change was decided at the union's 1980 national conference. But as a colleague and I observed (Frenkel and Coolican, 1984: 124) 'this issue was cast largely in terms of the anticipated impact of new technology in the metal industry rather than [in terms of] its current consequences'. We noted further that 'at the following national conference in 1982 there were few signs that technological change had made a substantial impact on employment or industrial relations in the metal industry.' However, at the most recent (1986) national conference, the impact of new technology on skill levels was addressed. Conference decided to establish a task-force to investigate this and related matters. In addition, new policies on improved trade training, multiskilling and upskilling of metalworkers were endorsed. Nevertheless, technological change has tended to be gradual and relatively uncontroversial, an experience very different from industries like printing and telecommunications.

References

Amalgamated Metal Workers' Union. 1985. 'Results of a Survey on Industrial Democracy'. mimeo.

Australian Bureau of Statistics, Catalogue Nos 6301.0; 6302.0; 6330.0; 8202.0 and 8307.0.

Australian Science and Technology Committee. 1985. *Computer Related Technologies in the Metal Trades Industry*. Canberra: AGPS.

Committee of Inquiry into Labour Market Programs (The Kirby Report). 1985. Canberra: AGPS.

Department of Employment and Industrial Relations. 1985. 'Review of the Labour Market for the Trades.' mimeo.

Frenkel, S. J. 1985. 'Managing Through the Recession: An Analysis of Employee Relations in the Engineering Industry.' mimeo.

Frenkel, S. J., and A. Coolican. 1984. *Unions Against Capitalism*. Sydney: Allen & Unwin.

Metal Trades Industry Association. 1984. 'National Survey of the Metal and Engineering Industry.'. mimeo.

Metal Trades Industry Association. 1985. 'National Survey of the Metal and Engineering Industry.' mimeo.

Schonberger, R. J. 1982. *Japanese Manufacturing Techniques*. New York: Free Press.

Part V

Technological Innovation and Workplace Relations

15

Information, Consultation and the Control of New Technology

Robert Price

'Innovation by negotiation' summarizes the general policy stance of British trade unions towards the application of microelectronic technology. From the TUC statement *Employment and Technology* (TUC 1979), through the programmes and policy statements of virtually every large union, a common theme can be identified of accepting the economic necessity of embracing the new technologies, while seeking to control their implementation. The formal focus of union policy has been on the demand for 'new technology agreements', designed to establish the employers' procedural commitment to the negotiation of change, and the sharing of the derived benefits between employers and employees.

The stated objectives of the unions were ambitious, emerging as they did from a movement strongly wedded to the post-Bullock scenario of extending the limits of collective bargaining into strategic areas of management decision-making. New technology agreements should provide the vehicle for a comprehensive joint regulation of the process of technical change; all change was to be by agreement, full information disclosure should take place, unions should be involved in systems design, and joint review procedures should be established. Additionally, there should be substantive agreements to protect jobs, reduce working time, provide retraining and strict health safeguards. The aspiration was for agreements to be 'sufficiently clear, comprehensive and accessible to allow the process of technical change to take place continuously and beneficially' (TUC, 1979).

It is generally accepted, however, that actual practice has diverged very substantially from aspiration. The survey evidence of Williams and Steward (1985), Hillage *et al.* (1986), Rathkey *et al.* (1982), the

Labour Research Department (1982) and NEDO (1983) tends strongly to the common conclusion that, with or without new technology agreements, British unions have exercised only very modest influence over new technology innovations. There have certainly been some improvements in information flows to workers and unions, and consultation with workers and their representatives has generally taken place. But these information and consultation processes have not led to an enhanced ability or desire to control the application of new technology. Instead, the response has been adaptive and accommodative, focusing on the defence of existing jobs and on the health aspects of new equipment. This picture is substantially confirmed by the Policy Studies Institute survey of 776 microelectronic user establishments, in which only 7 per cent cited opposition from unions as a problem faced in the introduction of the new technology. The clear implication of the PSI study (which is by far the largest of all the surveys carried out in the UK), is that unions have generally not sought to oppose managerially determined processes of technological change, provided that no compulsory redundancies were declared and that there was a reasonable attempt to inform and consult on proposed changes. 'The past record shows little underlying opposition by unions and the workforce to new technology, and often positive enthusiasm for it. . . . There is no evidence yet that these basically positive attitudes are changing' (Northcott *et al.*, 1985: 64). In particular, it seems incontrovertible that British unions have failed to tackle the underlying issues of job design, work organization and quality of work that might have allowed them to counterpose employer proposals with their own vision of new technology applications (Williams and Steward, 1984; Moore *et al.*, 1984).

An Alternative Perspective?

Despite the apparent unanimity of the survey evidence, a rather more positive perspective on the capacity of unions to exert influence on technological changes, emerges from some of the more detailed case studies that have been carried out in the past few years. Moore and Levie (1981: 2) found that the staff unions at GEC in Coventry had 'established a code for the introduction and use of VDUs which gives them some control over the way VDUs can be used'. They also cite the influence that shop stewards were able to exert over the equipment, layout and work organization of BL's Metro line (see also Francis, 1986). Wilkinson (1983), despite strong criticism of traditional methods of bargaining as 'wholly inadequate', also pointed to the capacity of many of the workers and unions in his sample to achieve important changes in working practices surrounding new technology by on-going processes of negotiation and renegotiation during the implementation and debugging of the technology'. Storey (1984), in a study of computerization in the insurance industry, found that relatively compliant employee attitudes and weak new technology agreements were to be found in conjunction with a 'failure fully to exploit the control potential of the technology. In large measure these companies have soft-pedalled' (1984: 44). This was explained in terms of the companies' desire to maintain a co-operative and harmonious set of relationships with staff and unions. Even weak unionism could produce a pre-emptive 'softening' of the hard edge of Taylorist work reorganization principles.

Finally, and most suggestively, Rose and Jones (in Knights *et al.*, 1985), in a series of six case studies, found substantial variations in the capacity of unions at plant level to resist management attempts to reorganize work and to bargain for non-pecuniary concessions in return for the introduction of new technology and other forms of flexibility. Their findings are, in an important sense, the obverse of the commonly accepted view that unions have continued to bargain about new technology in their traditional style; Rose and Jones found managements too, operating within the traditions and legacies of past industrial relations practices. Rather than using the recession and unions' lack of initiative in response to new technologies as a signal for a wholesale assault on work organization, managements were 'still prepared, indeed often deem it necessary, to elicit co-operation from unions with differing degrees of consultation and participation in the implementation of change' (1985: 99). Thus traditional forms of sectional and particularistic bargaining could achieve significant improvements in job content, gradings, training and job security.

An 'alternative' perspective on British unions' ability to defend their members' interests in the introduction of new technology might then involve a downwards revision of the ambitious 'targets' built into programmatic statements such as the TUC's *Employment and Technology* document, and a closer look at the practice of office and shopfloor unionism where incremental and often small-scale changes are taking place. Research in eight manual worker environments in the plastics industry, and four white-collar environments in engineering, supports the perspective adopted by Rose and Jones, and some of the conclusions cited above. In none of these cases had the introduction of new technologies caused a significant rupture from past industrial relations practices. Instead, technological change was handled in 'traditional' ways, with considerable success for both managements and unions, suggesting that the apparent lack of opposition experienced by British managements to new technology should not be taken automatically to imply union acquiescence in management plans. The following section reviews the broad pattern of bargaining relationships which emerged from these studies and proposes a number of reasons for the variance between this evidence and the standard 'model' of union ineffectiveness outlined in the introduction.

Evidence from the Case Studies

The 12 cases cover a wide range of production environments and plant size. They range from a specialist metals research organization with less than 100 white-collar employees and a company producing a semi-finished plastic foam material with 100 manual and 60 white-collar workers, to a large building products firm with over 600 manual workers, and the central administrative organization of an automotive parts producer with some 800 white-collar workers. The manual worker environments were selected to reflect 'good practice' in information provision to unions and employees, and the white-collar environments primarily because they represented, for the unions concerned, well-organized workforces with good track records in workplace bargaining. They were studied between April 1983 and April 1985, as part of two separate research projects.

Detailed accounts of the case studies cannot, of course, be provided here, and much of the detail would in any event not be relevant to the argument of this chapter. This is that the principal factor affecting the capacity of workers to influence the introduction of new technologies was the pattern of union–management relationships built up in the decade (or more) of bargaining experience prior to the current period of recession and technological innovation. The existence or absence of an information agreement or a new technology agreement, and the precise terms of such agreements, were found to be less important in practice than the set of customs, practices and understandings with which key negotiators on either side approached the bargaining issues involved.

To illustrate this proposition, the sample can be subdivided into two groups: eight cases of 'co-operative' industrial relations traditions, and four cases of 'conflictual' past relationships. The 'co-operative' group was characterized by a history of stable inter-union and management–union relationships, some form of consultative system alongside grievance and bargaining machinery, and considerable autonomy in industrial relations from 'external' management and union involvement. Beyond these common basic elements, the eight firms varied widely in other important respects. Half of this group had suffered very large cuts in employment between 1980 and the date of the research; the others had experienced only minor employment reductions. Three of the plants had substantial numbers of workers of ethnic minority origin; only two of the white-collar groups were in this category, but they were respectively dominated by female and male employees. Six of the plants had some form of information 'agreement' of varying degrees of formality; only two had formal new technology agreements – perhaps predictably, the two white-collar cases. The technology used in these eight plants varied widely, from injection moulding machines and polyethylene fabrication plant, to the conventional design and administrative equipment of the 'old technology' office. As will be seen below, this group of plants also varied widely in the scale of their experiences with new technology; some had gone through rapid and major changes, others had experienced only piecemeal and minor changes. In every case, however, the introduction of new technology had been handled in a way which permitted a significant role for workplace union representation, and which produced outcomes that were clearly marked by union influence.

The 'conflictual' group also displayed substantial internal heterogeneity in respect of product, technology, size, exposure to competitive pressure and labour force composition. Their common feature was a lengthy history of mutual antagonism between unions and management. Inter-union relationships in these plants also tended to be tense and occasionally hostile; this would typically involve craft/process worker union antagonism or manual/white-collar union tensions. This strand of mutual antagonism did not necessarily entail overt industrial conflict. The feelings of antagonism were reflected in dismissive or aggressive statements about management–union behaviour, frequent use of the grievance and disputes machinery, little or no consultative machinery and constant disagreements about union rights (for example, on time off or on health and safety matters), and management's 'style' of management. This pattern of mutual distrust and antagonism was reflected in the handling of technological innovation. While the scale of such innovation varied, the common themes (throughout the four cases) were the attempt to minimize union influence and an unwillingness to consider alternative outcomes to those proposed by management.

'Co-operative' Workplace Relationships – Information and Consultation in Practice

The formal structures of negotiation and consultation in the plants in the 'co-operative' group were, of course, very varied. The common feature, however, as indicated earlier, was the existence of a reasonably long-standing consultative system or body, alongside the disputes and bargaining machinery. Most of these systems dated from the early or mid-1970s. In a middle-sized company producing polyethylene products for the civil engineering sector, for example, there was a quarterly joint consultative committee at which employee representatives elected from workforce constituencies, together with the four senior stewards, met the chief executive to discuss company strategy and financial position. In addition the full joint shop stewards committee, representing TGWU, AEU and EETPU, had bi-monthly 'information' meetings with the manufacturing director and the personnel manager. In a rather bigger plant making a range of mouldings in thermoplastic and rubber, a monthly works council meeting brought together all twelve manual senior stewards and four senior managers, including the plant managing director, for a general 'information exchange'. In practice, these sessions had evolved over the years into wide-ranging 'predictive bargaining' meetings, in which both sides floated ideas about future developments on a 'without prejudice' basis. In the smaller of the two white-collar cases in the 'co-operative' group (a metals research organization), a formal information agreement led to monthly consultative meetings on the organization's development and planning objectives.

Similar information and consultation systems were to be found in the other plants in the 'co-operative' group; but rather than extend this set of illustrative descriptions, the question of effectiveness should be addressed. What did these arrangements actually mean in practice? Were the union representatives able to use them in the interests of their members, or were they simply talking shops in which the managements were able to persuade the stewards to see things their way? Generalized answers to these questions are difficult, since the range of experience is wide. Nevertheless, interviews conducted with both managements and stewards produced strongly positive views as to the value of their information and consultation arrangements during the introduction of new technology in their plants. Three examples will illustrate this.

In a medium-sized injection moulding plant, making parts for the highly competitive automotive and food industries, a new management team had begun to develop computer-controlled monitoring and control systems for the moulding machines and a computerized stock control and warehousing system. The choice of machinery had been management's alone. But the information and consultation process, starting as soon as the decision to install the new machinery had been made, was comprehensive and intense. The whole workforce (process workers, indirects and maintenance) was organized by the AEU, which may have eased the strains between the skilled and non-skilled groups that emerged as the new machinery was discussed. The two main implications of the new systems were clearly identified as: (a) a potential loss of 10 to 15 per cent of jobs, concentrated amongst the process workers; and (b) the need for substantial retraining and reorganization of the manufacturing, maintenance and quality control functions. As the manufacturing director put it: 'by the end of this process, 80 per cent of the jobs in this plant will be either non-existent or reorganized'. The union's initial objectives were to guarantee no compulsory redundancies or reductions in pay levels; in short, the traditional defensive reactions of British shopfloor organization. The management, however, as well as taking the information and consultation process seriously, was also anxious to involve the workforce fully in the changes. Consequently, they established a series of shopfloor working groups (with one of the stewards always present) in which the restructuring was discussed in great detail. Both stewards and management agreed that their ideas and approach to the new systems changed radically as a consequence of these group discussions. The stewards pressed increasingly for a revision of job descriptions which straddled the traditional craft/process boundary, and the management began to take retraining increasingly seriously. The outcome was a reorganization of production and warehousing facilities that was substantially different from the concept with which management had begun the consultation process. Process and warehousing workers were retrained to take wider quality control and computer control responsibilities; foremen and supervisors were redeployed in technical and administrative functions. At the time of the research there had been no compulsory redundancies, and management admitted that this concession to the unions meant that potential cost savings were taking longer to achieve than they would have wished. Soon after, some 20 of the (non-unionized) clerical employees were declared redundant – very clearly the casualties of the shopfloor's success in achieving an innovative reorganization of work.

A second example is a firm producing decorative laminates, employing some 600 people, and virtually 100 per cent unionized in all non-managerial grades. Technical change had been more evolutionary and piecemeal than in the previous case, but, here too, changes were introduced via an elaborate and multi-faceted system of consultation and negotiation. From 1979 to the date of the research in 1984, there had been a gradual shift towards computer-controlled production and warehousing systems. Throughout the process, detailed information had been provided on the nature of the new equipment, and new working practices were elaborated in a series of meetings involving the workers concerned and a joint policy committee of union and management representatives. Management claimed that the precise form of work reorganization had been so much influenced by the consultation/discussion process that it was impossible to say how far it differed from what might have emerged without consultation. Union representatives considered that they had strongly influenced the layout of the plant, job allocations and the reorganization of the maintenance functions to involve 'multiskilling' and craft/process worker flexibility.

A third, very simple, illustration comes from the white-collar case studies. A small specialist engineering company wanted to introduce word processing into its clerical/secretarial functions; some 30 employees, members of APEX, were involved. The company had no clear idea of the type of system that they should choose, nor what the precise benefits might be. From the outset, they involved the APEX steward in the various presentations by equipment sales-people, and asked him to advise on the 'best' way of changing over to word processing. On advice from APEX officers, the steward drew up and agreed a new technology agreement which incorporated a range of safeguards on both health and safety, and job design. The final outcome was a system incorporating union 'best practice' advice, although without any substantive concessions on pay and hours.

To summarize broadly from these examples and the wider group of eight cases which they represent, three conclusions can be drawn. First, it is evident that the kind of questionnaire data on which the more pessimistic assessments of union ability to influence new technologies have been based, are unsuitable for tapping the complexities of the union–management relationships located in dynamic and plant-specific sets of customs and practices. In each of the three examples quoted briefly above, it is unlikely that a senior manager answering a questionnaire would have said that the introduction of new technology had been made problematic by the activities of the unions; nor would they have claimed that their new technology and information agreements, where they existed, had posed significant restraints on the achievement of managerial objectives. But while these answers might have suggested union acquiescence and simple job defensiveness, the reality was far more complex.

Secondly, new technologies obviously emerge into established bargaining environments. Since the early 1970s, there has been increased interest in linking bargaining with information and consultation procedures (Hawes and Brookes, 1980: 333–61; but see also MacInnes, 1985). Thus even where no formal information disclosure agreement exists – as in several of these cases – informal assumptions and modes of behaviour have developed which put a premium on good information flows, and a consultative/participative style of management. It is completely unremarkable in this type of climate for new technology to be handled in the same way, particularly where their introduction is on a fairly gradual and piecemeal basis.

Thirdly, the consequence of a relatively relaxed and mutually non-antagonistic industrial relations environment, coupled with relatively good information flows and often quite elaborate consultation systems, is that the process of debate and discussion about new technology can lead to a wide degree of agreement. It is then very difficult to ascribe the outcome to specifically management or union objectives. The evidence of the case studies was that management proposed, but that consultation disposed.

Despite a generally positive evaluation of the ability of the unions in these eight cases to use information and consultation rights to further employee interests, two qualifications should be entered here for further discussion below. It will be clear from the examples presented above that to a very important degree, the systems of information and consultation were management sponsored. That is, they formed part of a gradually evolving management 'style' in the industrial relations and personnel sphere. For many of the small and medium-sized plants studied here, the 'sophisticated modern' style of management had become a well established feature of boardroom policy (Bain, 1983: 113). This was frequently reinforced by the experiences of the recession and the awareness of the scale of changes that were likely to result from new technology. Rather than abandon their approach in a period of stress, managements increased their emphasis on 'taking the workforce along with them'. The managements in this sample were committed to a genuine consultative style because they saw it as in their interests, and because their experience of this style had been positive in the past. This leads to the second observation: none of the workplace union organizations studied here had challenged management's right to set the parameters of the debate about new technology, or posed clear-cut alternatives to management's proposals. While the negotiation and consultation process certainly modified the terms of implementation, the outcomes were all within the bounds of managerially defined 'acceptability'.

'Conflictual' Workplace Relationships – the Imposition of New Technology

The four plants categorized as exhibiting a 'conflictual' style of workplace industrial relations had had markedly different experiences in the introduction of new technology. The common element in the development of industrial relations in these firms was an oscillation between 'constitutionalist' and 'standard modern' approaches, reflecting 'complex and shifting blends of unitary and pluralistic perspectives' (Fox, 1974: 308; Bain, 1983: 115–16). A plant producing plastic building products with over 600 employees was typical of this style of relationship. There was a lawful closed shop for manual workers, and the union organization in the plant was well resourced with office, telephone and flexible 'time off' procedures. Management was organized in a relatively rigid hierarchical and bureaucratic fashion, and clung firmly to the notion of managerial prerogatives and the need to defend them from what was seen as an aggressive union continually seeking to encroach into the 'secret garden'. All formal management–union contact was within the grievance and disputes machinery; informal conversations and phone contact tended to relate to issues that would potentially, or had already, gone into the machinery. There was no forum for employee and management representatives to meet regularly to discuss general company policy and future plans and developments; they would grudgingly accept each other's legitimacy, but lacked the mutual respect and trust needed to advance beyond the conflictual, distributive aspects of collective bargaining. Management criticized the stewards for conveying partial and biased information about negotiations to their members, thereby not doing justice to the firm's case. The stewards criticized management vigorously for failing to provide information, for distrusting the union, and for not seeking to stimulate union or employee input to company decisions.

This pattern of mistrust and hostility was neatly exemplified by the introduction of a new production facility site using computer control, and associated changes in machine manning and plant layout in the older buildings on the site. There was no consultation on the design of the new facility, nor on the type of jobs to be created in it. Negotiations on the changes in the existing buildings took place over a couple of months against a background of an assurance that staff reductions would be handled by natural wastage and early retirements. At the end of this period, with agreement still some way off, the company declared 50 redundancies. Not surprisingly, the union representatives felt that they had failed to influence management; but despite the apparent strength of the shopfloor organization, they feared that a call for industrial action would not be popular and well supported. The 50 redundancies were obtained voluntarily and the new facility was opened on management's terms.

In similar vein, a complete on-line computer system to control stocks, ordering, invoicing and warehousing was installed in the largest of the white-collar environments studied, with merely perfunctory consultation well after all the important decisions had been taken. This too was a well organized plant, but with a long history of weak union representatives and hawkish management. They considered that the organizational weakness of the union, despite strength in numbers, justified their 'arms length' approach. 'Why should *we* take them seriously, when they don't take themselves seriously?' said one line manager. A new technology agreement had been signed in this plant, but when it came to the implementation of the system, management claimed that they could not be bound by the strict terms of an agreement signed three years earlier in a different economic climate. The company ran a series of 'propaganda' sessions about the new system, but undertook no consultation or bargaining about the precise implications of the system for individual jobs. The reduced staffing levels associated with the new equipment were achieved voluntarily, thus keeping the company within the terms of the new technology agreement, and, in the view of the stewards, removing the only issue that might have stimulated concerted opposition among the workforce.

There are, thus, marked contrasts between these two cases. But the common features are striking: high levels of union density but without a strong bargaining relationship; mutual distrust and often contempt between management and union representatives; no joint consultative forum, or established system of management–union–employee communications. In both plants, as in the others in the 'conflictual' group, managements were consciously influenced in their dealings with the unions over new technology by the belief that the economic climate had weakened employee willingness to resist, and that they should take advantage of this situation to achieve objectives quickly and effectively. Compared to the 'co-operative' plants, technological change in this group *was* implemented quickly, and with little or no deviation from the planned schedule of change. In both co-operative and conflictual groups, the pre-existing patterns of institutions, styles of interaction and sets of understandings seemed to determine the approach to the introduction of new technology.

Concluding Comments

These case studies were carried out in environments where unions might not have been expected to exercise significant influence on the introduction of new technologies. Plastics companies are not noted for their history of vigorous shopfloor organization; and white-collar (clerical and technical) employees are also not generally associated with a history of control struggles. Indeed, with one exception, the plants studied were either non-unionized or not in existence as recently as 1965. However, in two-thirds of the cases, union influence on work organization following a 'new technology' innovation was significant and recognized as such by management. This influence derived, not from the chosen vehicle of the trade union movement, the new technology agreement, although some of these did exist, but from more general industrial relations procedures and understandings that underpinned union–management relationships in these plants. These understandings involved a relatively open approach to information provision, albeit after management had defined the general nature of the technology to be installed; a commitment to a variety of consultation procedures, usually involving a blend of union and employee-specific channels; and a personnel/industrial relations philosophy or style which emphasized employee 'commitment' to the enterprise.

Conversely, plants characterized by an absence of information and consultation channels of at least moderate intensity, and a history of management–union antagonism, were unlikely to show significant union influence on new technology-related work organization, even where formal new technology agreements existed.

By way of conclusion, four interrelated propositions can be advanced arising out of the evidence presented here and the wider spectrum of evidence on union involvement in the introduction of new technology.

1. There is no inconsistency between this case study evidence and the large-scale survey evidence on one key issue. Overwhelmingly, British union involvement in work reorganization consequential on new technology takes place *after* the initial planning stages have been carried out by management. The ambitious scenario mapped out by the TUC whereby unions would pose fundamental questions about the social role of technology and intervene pre-emptively in job design, has simply not been on the agenda. Nevertheless, involvement in the implementation process alone can still yield important benefits for employees.

2. The case study evidence reflects a growing emphasis on consultation and information procedures over the past decade, alongside existing bargaining and grievance procedures. These consultation systems are generally designed to complement and supplement the 'distributive' character of collective bargaining, and have become linked in many cases with quite explicit management policies to encourage employee commitment to the enterprise. While these systems are manifestly employer sponsored, it seems clear that if they are to have any credibility, they have to be seen to work effectively. They thus acquire a dynamic and impetus of their own. While they might have been initiated in a fairly crude attempt to co-opt or manipulate, their operation results in a 'virtuous circle' of gains in mutual confidence, and more positive attitudes to joint decision-making.

3. In the specific case of new technology, a further factor seems to have been important in reinforcing the willingness to consult and inform unions and employees. The uncertainty which many employers themselves feel about what to do with microprocessor-based technology, and the awareness that a positive attitude from the workforce is probably crucial if new equipment and new processes are to produce positive benefits, have made consultation almost a necessity to get new systems off the ground successfully. As Rose and Jones put it, 'managements are still prepared, indeed *often deem it necessary* to elicit co-operation from union representatives in the implementation of change' (Knights *et al.*, 1985: 99; my emphasis).

4. Union–employee influence on work reorganization, through both bargaining and consultative channels, has been limited in its *character*. Unions have simply not begun to tackle the issues inherent in the notion of 'worker-oriented design' as proposed by some German unions

(IG Metall, 1984) and by some union strategists in the United Kingdom. Thus while it may not be totally inaccurate to speak of genuine areas of joint implementation of new technology, this must be understood to imply implementation within the acceptable limits for managements.

While this chapter has sought to illustrate the weaknesses of the standard analysis of union effectiveness in the implementation of new technology, it would be naive to ignore the reality that if unions are to move out of an essentially defensive and reactive posture, far greater resources and much greater co-ordination of those resources across unions are necessary. Neither of these changes seems likely in the near future – and 'new' technology will soon be old technology!

Note

The author is grateful to Stephen Wood of the London School of Economics and Rod Martin of Imperial College for helpful comments on this chapter.

References

Bain, G. (ed.) 1983. *Industrial Relations in Britain*. Oxford: Basil Blackwell.

Fox, A. 1974. *Beyond Contract: Work, Power and Trust Relations*. London: Faber.

Francis, A. 1986. *New Technology at Work*. Oxford: Oxford University Press.

Hawes, W., and C. Brookes. 1980. 'Change and Research: Joint Consultation in Industry'. *Employment Gazette*.

Hillage, J., N. Meager and A. Rajan. 1986. *Technology Agreements in Practice: the Experience So Far*. IMS Report No. 113.

IG Metall. 1984. *Der Mensch Muss Bleiben: Arbeitsprogramm Arbeit und Technik*.

Knights, D., H. Willmott, and D. Collinson. (ed.) 1985. *Job Redesign: Critical Perspectives on the Labour Process*. Aldershot: Gower.

Labour Research Department. 1982. *Survey of New Technology*. Bargaining Report No. 22.

MacInnes, J. 1985. 'Conjuring Up Consultation: The Role and Extent of Joint Consultation in Post-War Private Manufacturing Industry'. *British Journal of Industrial Relations*. March.

Moore, R., and H. Levie. 1981. *New Technology and Trade Union Organisation*. Oxford: TURU Occasional Papers No. 71.

——., H. Levie *et al.* 1984. *The Control of Frontiers: Workers and New Technology, Disclosure and Use of Company Information*. Oxford: Ruskin College.

NEDO. 1983. *The Introduction of New Technology*. London: HMSO.

Northcott, J., M. Fogarty, and M. Trevor. 1985. *Chips and Jobs: Acceptance of Technology at Work*. London: PSI.

Rathkey, P., W. Fricke, and P. Konig. 1982. *New Technology and Changes: An Anglo-German Comparison*. Newcastle: Conway Foundation.

Storey, J. 1984. *The Phoney War? New Office Technology: Organisation and Control*. mimeo.

TUC. 1979. *Employment and Technology*. London: TUC.

Wilkinson, B. 1983. *The Shopfloor Politics of New Technology*. London: Heinemann.

Williams, R., and F. Steward. 1984. *The Role of the Parties Concerned in the Introduction of New Technology*. Dublin: European Foundation for the Improvement of Living and Working Conditions.

——., and F. Steward. 1985. 'Technology Agreements in Great Britain: A Survey 1977–83', *Industrial Relations Journal*. Autumn.

16

Consensual Adaptation to New Technology: The Finnish Case

Pertti Koistinen and Kari Lilja

Introduction

It can be assumed that the implementation of new technology in work organizations is such a critical incident that it reveals important aspects of the economic, social, political and ideological contexts of the immediate labour process. Thus in this chapter we are not directly interested in new technology and its social implications; instead, we try to outline some national characteristics of the implementation of new technology in Finland.

There are rather persistent economic and institutional structures in Finland which produce a nationally specific context for implementing new technology. Except in very narrow fields the Finnish economy adapts only to technological innovation produced in countries with larger economic resources. Thus the technological strategy of Finnish firms has been based on the rapid initiation of the most advanced practice available. This has saved research and development costs. It has only required an up-to-date educational system which produces qualifications to monitor and forecast trends in technological development on a wide scale (Koistinen, 1985a).

Though it is the companies and their management which play the active role in the development and implementation of new technology we focus this chapter on the role of the workers in manufacturing industries. In general, it can be argued that workers and the trade union movement have supported the technological strategy of Finnish capital and accepted it as a national necessity. This has contributed to the rapid implementation of new technology at the level of individual work organizations. The costs of the organizational and occupational changes have been carried mainly by individuals and families, to some extent by paternalistic firms until the 1960s and since then increasingly by the welfare state.

Thus our starting point, when looking at the role of the workers, is a phenomenon which could be called the consensual adaptation to the implementation of new technology. This contrasts radically with an oppositional response based on a tradition of job control (including for instance unilateral regulation of the standards of work by the members of the craft, clear demarcations between job categories and restrictive practices) (cf. Hyman and Elger, 1981).

The purpose of this chapter is to elaborate further the idea of a consensual adaptation to the introduction of new technologies in the Finnish context. We illustrate the phenomenon with observations from fieldwork in three modern engineering plants and provide an historical and institutional framework to explain the phenomenon.[1]

The Underlying Logic of Consensual Adaptation

It is quite easy to understand why the introduction of modern industry has been welcome to the working class it has created in Finland. The alternative sources of livelihood have been based on agriculture and on forestry. As late as 1950, 40 per cent of the employed population was working in the primary sector. The erection of new factories opened opportunities to wage-labour which have been viewed as an advancement in the status hierarchy of a predominantly rural society. The employment share of manufacturing industries expanded until 1974, and in the metal industries the number of workers was still growing in 1982. Taken as a whole, the change in the Finnish occupational structure since 1950 has been one of the most dramatic among the OECD countries. This is the broad background for the arguments grounding the phenomenon of consensual adaptation to the implementation of new technology. The structural change would not have been possible without a mentality in favour of its implementation.

The underlying logic of consensual adaptation is based on a positive-sum game where the exchange relationship – the implicit contract – gives even the unprivileged a relative advantage over his or her next best alternative solution. The long period of extensive growth in manufacturing industries directed the attention of the trade union movement mainly to the problems of income distribution and welfare policies, and few regulations were imposed on managerial prerogatives in the organization of production. At the same time the extensive industrial expansion left plenty of room for individual choice in the labour market, and the phenomenon of self-selection to jobs could clearly be detected. This resulted in great migration within the country and even abroad, mainly to Sweden. This has undermined the development of collectivistic traditions when facing organizational and occupational changes.

Now that extensive growth of manufacturing industries has stopped even in Finland, it must be asked, is there still a material basis for a positive-sum game? It is clear that there is a growing tendency for workers in the labour market to become labelled as winners or losers, due to rationalization in the basic industries. This means that within the working class the positive-sum game does not function any more; but this may even strengthen consensual adaptation. The fear of becoming a loser in the labour market reinforces the individualistic approach; workers accept the need to maintain the international competitiveness of Finnish industries and competition in the labour market has increased.

The Case of Young Workers in the Engineering Plants

To take a closer look at consensual adaptation in the present situation we report some observations from fieldwork conducted in three modern engineering plants. The plants are leaders in the introduction of modern CAD/CAM technology. All plants belong to divisions of large Finnish corporations which have subsidiaries also in other countries. During the first phase of the fieldwork 80 workers were interviewed (see Koistinen, 1985b). The interviews reveal that a large proportion of semiskilled workers view the introduction of new technology as rewarding. Their work has changed substantially, and they have been able to acquire new skills through different types of training. This has improved their long-term labour market position. While employed in one of the leading plants in their local labour market they do not intend to change their place of employment.

With the introduction of new technology, considerable changes have occurred in supervisory practices. The workers involved with new technology have direct access to engineers whom they can consult when problems in their work occur. They can short-circuit the immediate foremen who do not have superior knowledge in the application of new technology. This gives the workers in question a feeling that they have more autonomy in their work than before. This is strengthened by the tacit skills related to their concrete tasks, which make even the engineers dependent on their contribution to the planning process.

This implicit contract between the engineers and a network of workers has at its core the underlying consensus that both groups are committed to the competitive race to implement new technology in the production process. It is assumed that the cost efficiencies created by this implementation process will be beneficial to both groups, leading to increased job security, higher wages and salaries, career prospects and opportunities for training.

In all the plants the group involved in this implicit contract is a set of young male workers with intrinsic interest in their work. They have become familiar with EDP technology not only through their training on the job but also through hobbies linked to microelectronics and home computers. A further social condition for this contract has been the respect which the young workers show to those specialists who by their broader education can give advice in complicated situations. This has further accentuated the emerging high trust relation.

How has it been possible to accommodate the results of consensual adaptation into the social system of the plants? There are clearly segments of the labour force whose jobs have not changed and who consider the implementation of new technology a threat to their existing job contents. Such is the case especially among unskilled workers, and also older workers complain about the degradation of their traditional skills. Transfers to new jobs have, however, been based on voluntary changes when departments or phases of the work flow have been converted to new technology. Older workers tend to continue to use old machines and production methods. Moreover, the workers do not evaluate their position in the social system of the workplace only according to the characteristics of their work. Their evaluations are affected by the social qualities of the immediate work group and other social identities. For instance, irrespective of their skill level and specific occupation all workers employed by the leading plants in the local labour market acquire a sense of status and self-esteem.

The social system of the plant is also mediated by the mechanism of workers' representation and participation. This mechanism is dominated by middle-aged and older workers. While traditional issues concerning terms of employment and labour protection have retained their importance and disputes are processed through that mechanism, the established elite of the workers do not feel themselves pushed aside. The young workers who have contributed to consensual adaptation belong to the union branch but do not participate in trade union activities. It appears also that the young workers do not find it difficult to pursue their work-related interests in their immediate supervisory context and in the participatory machinery (project groups, planning subgroups, etc.).

To summarize the immediate conditions of consensual adaptation we would emphasize the following points:

it appears that management had carefully selected workers for the new parts of the production system (young, semiskilled males with personal interest in new technology, for instance through leisure activities);

the workers selected for the new tasks are clearly better off compared to their earlier jobs and compared to the employment prospects of those who are not selected;

the workers and even foremen did not have an overview of the rather dramatic change process which was going on with the introduction of new methods, equipment and products. For this reason it was also impossible for them to develop joint interpretations of the situation, and it cannot be assumed that the different categories of workers, technicians and foremen would have consistent preferences and non-conflicting interests in the implementation process of new technology;

the change process occurs in a sheltered niche from the point of view of the employment situation of the workers. We are dealing with workers who are winners in the labour market.

But there are also wider historical and institutional conditions for consensual adaptation, and they are discussed below.

Societal Influences on Consensual Adaptation

The late take-off of industrialization and its narrow base in Finland until the 1960s explains why we encounter decade after decade workers who are the first generation in the family employed in manufacturing industries. This means that socialization to specific jobs or crafts has rarely started at home. Such first-generation workers have proved to be quite flexible: they have been ready to accept different types of jobs and it takes time before workers develop regulatory practices for the use of their labour power. Moreover, the first-generation workers who have migrated to towns from rural areas have not lost contact with the mentality of working hard in order to earn one's living.

Clear occupational boundaries are also avoided in the recruitment practices of the firms. Even in the metal industry where the labour process is identified normally through a variety of craft identities, it has been common practice in Finland to hire workers more to the internal labour markets than to specific jobs. This is partly due to the small size of machine shops, the predominance of small-batch production and the low level of specialization. These have been conditions for small-group organization and flexible use of labour.

While occupational cultures have been rare and do not constitute an important element in social networks, similarly there is a lack of long traditions of trade unionism. Before the Second World War the role of trade unions was negligible. The system of collective bargaining was established on a permanent basis at industry level in the latter part of the 1940s. Since then the unions have been able to influence the terms of employment, but there have been very few attempts to influence the content of jobs and other aspects of work organization. In these respects the Finnish unions depart clearly from the Scandinavian model where trade unions have had legitimate status in the eyes of employers for a long time.[2]

While the Finnish trade union movement is organized on an industrial basis, union structures do not support workers' occupational identities. Craft unions have been exceptions in manufacturing, the only significant case being the electricians' union which is becoming increasingly based in energy production. When new technology has caused changes in the established structures of occupations the unions have not been in danger of losing their foothold in an industry or a workplace: the demarcation disputes have taken place within the same union. Thus unions have been willing to support the modernization efforts of management and resistance to change has not become a social movement beyond the level of individual workplaces.

It is clear that the Finnish trade union movement has little capacity to deal with managerial issues. This is accentuated during the present rapid structural change in the economy. Since the end of the 1960s there have existed procedural mechanisms, and since 1979 a law which requires information disclosure and advance negotiations when significant changes are made at the workplace. These have created a new constitutional basis for workplace industrial relations; but it is clear that the trade union movement has failed to establish new custom and practice rules around these procedural reforms.

So far the argument has been that union bargaining power in Finland has rarely been used to resist technological change. We can qualify this statement by considering an exceptional case: the pulp and paper industry. It represents the industrial core of the Finnish economy with relatively large mills and long industrial traditions. The particularistic cultural aspects are supported by isolated mill communities very much according to the well-known Kerr–Siegel (1954) hypothesis. Within these communities strong industrial and occupational traditions have emerged and been transmitted from one generation to another. Due to the simple structure of the industry the workers have been able to create a strong industrial union. The Paper Workers' Union has been able to negotiate a clause in its collective bargaining contract which enables workers' representatives at local level to exert considerable influence on the implementation of new technology and on other changes in conditions of work.

The contract stipulates that a local agreement must be reached on wages to be paid before a major change in the working conditions is implemented. This is an obscure statement because it is a compromise which was very reluctantly agreed by the employers. They have not been willing to make agreements on issues which they consider to be their prerogative. For this reason the clause appears as a wage clause and it must be admitted that it is mainly used as a way to get extra increases in wages. But this paragraph gives workers' representatives considerable power to influence the shaping of the working conditions when, for instance, new technology is introduced. In practice workers' representatives are involved in extensive project work when major changes are made. The implementation of the changes can be delayed by the senior shop steward if no wage agreement is reached, and this possibility puts considerable pressure on management to meet the demands of the workers, at least in monetary terms. The clause has, however, not retarded technological development in the Finnish paper industry, which is internationally known for its high level of technological sophistication. Dramatic changes have been introduced in the organization of work and manning levels since 1968, when the clause was introduced into the industry-wide contract. Management has not compromised on this aspect, but technological changes have been bought by considerable wage increases which have raised the average wages of paper workers to the top of the manufacturing league table. Paper workers have also been the first to develop and press through qualitative demands with respect to their work and its content; the work environment in the control rooms bears little resemblance to traditional manufacturing work. But despite the strong bargaining position of the paper workers consensual adaptation has been the dominant approach in responding to managerial initiatives for technological change. This vindicates again the line of argument presented above.

It must be noted, however, that the characteristics of consensual adaptation appear to contrast with the history of Finnish industrial relations (Knoellinger, 1960; Lilja, 1983). This has involved intensive class conflict, and in a cross-national perspective the propensity for industrial action has been rated as very high (see, e.g., Korpi and Shalev, 1980). By the end of the 1960s the strike waves reached the workplace level in a decentralized fashion, demonstrating the existence of a strong rank-and-file movement.

The disparity between this picture and the notion of consensual adaptation can be explained by considering the motives of the industrial disputes. They have been linked mainly to wage claims and wage settlements. Though these may conceal the influence of suppressed political cleavages, the point is that workers have very seldom contested managerial authority in organizational issues. Such a trend has been prevalent even during the 1980s when large-scale redundancies and plant closures have occurred. The percentage of industrial action related to reductions in the labour force has been on average below 5 per cent during the 1980s (Statistical Office of Finland. TY 1985: 21).

This duality in the pattern of workplace industrial relations could be characterized as a generalized effect at the national level of the phenomenon, documented by Daniel (1973) in the social context of a single factory. He found that social relations in the factory were regulated by two types of climate: the conflict-laden bargaining climate and the co-operative climate of the immediate labour process. During intensive bargaining phases when there are problems in finding acceptable solutions, the bargaining climate dominates the social relations of the immediate labour process, leading to confrontation. But when the bargaining phase is over the climate becomes milder again.

It appears that Finnish management has been able to reactivate consensual adaptation, despite the periodic open conflicts in the sphere of distribution. It is clear that this has not occurred by using sophisticated human relations techniques: Finnish engineers have not been good at that. One hypothesis could be that it results from the high respect which engineers have enjoyed in the eyes of the craft workers due to their education and technical expertise. When asked what type of occupation the male craft workers would like to suggest for their sons they often reply: an engineer. This reflects the strong 'Technik' culture of Finnish manufacturing very much in line with the German tradition (cf. Child *et al.*, 1983).

Concluding Remarks

Our characterization of consensual adaptation to the implementation of technological innovation has implied a rather passive role for the workers. This need not be the case. Normally we come to know only the end result of a long planning and implementation process, and the struggles and negotiations included in it have left behind very few documents. It is known that different levels and functions of management have very seldom a coherent strategy to deal with the problems of new technology. This is quite natural given the great uncertainties and risks related to such investments. This leaves plenty of room for individual workers and occupational groups to bargain. The new organizational arrangements emerge in an incremental way from the experiments made, and the workers involved are in a good position to influence the ways in which tasks and jobs are defined both in the vertical and horizontal dimensions.

Notes

1. It is relevant to note that we restrict our discussion and generalization only to work cultures which are dominated by working-class males. Thus

from the gender perspective we are not covering the whole picture.
2. This was also the basis of joint projects to experiment with different types of work organization in the Scandinavian countries (see Sandberg, 1982).

References

Child, J., M. Fores, I. Glover, and P. A. Lawrence. 1983. 'A Price to Pay? Professionalism and Work Organisation in Britain and West Germany'. *Sociology*. Vol. 17, no. 1, 63–78.
Daniel, W. W. 1973. 'Understanding Employee Behaviour in its Context: Illustrations from Productivity Bargaining. *Man and Organization*. Ed. J. Child. London: Allen & Unwin.
Hyman, R., and A. Elger. 1981. 'Job Controls, the Employers' Offensive and Alternative Strategies'. *Capital and Class*. Vol. 15, 115–49.
Kerr, C. and A. Siegel. 1954. 'The Interindustry Propensity to Strike – An International Comparison'. *Industrial Conflict*. Ed. A. Kornhauser, R. Dubin and A. M. Ross. New York: McGraw-Hill.
Koistinen, P. 1985a. 'On the Determinants of Technological Development: Some Viewpoints on the Technological Development of the Finnish Industry'. *Acta Sociologica*. Vol. 28, no. 1, 3–20.
——. 1985b. 'The Use of Labour in Modern Plants of Engineering Industry: Some Conclusions of the Finnish Case-Studies'. University of Joensuu: Karelian Institute, Working Papers 5/1985.
Korpi, W., and M. Shalev. 1980. 'Strikes, Power, and Politics in the Western Nations, 1900–1976'. *Political Power and Social Theory*, Vol. 1, 301–34.
Knoellinger, C. E. 1960. *Labor in Finland*. Cambridge, Mass.: Harvard University Press.
Lilja, K. 1983. 'Workers' Workplace Organisations'. Helsinki School of Economics. Series A: 39. Helsinki.
Sandberg, T. 1982. *Work Organisation and Autonomous Groups*. LiberFörlag; Lund.
Statistical Office of Finland. 1985. TY 1985: 21.

17

Industrial Relations and Workers' Representation at Workplace Level in France

Sabine Erbès-Seguin

The period of the government of the Left (1981–6) was marked by major changes in work relations particularly at company level. These reflected not only short-term political and economic influences, but also longer-term trends dating back to the early 1950s. It is therefore necessary to give first a synopsis of the basic features of the industrial relations system and its evolution over the past 10–15 years, before describing in more detail recent developments at company and workplace levels.

Basic Trends in Industrial Relations in France

Between 1950 and the end of the 1970s, a period of continuous economic growth, a relatively institutionalized (though limited) system of sharing the benefits of growth was established. There seemed to be no 'technical' limits to negotiations on wages and fringe benefits, while economic expansion appeared permanent; and indeed many economists considered that the bargaining system facilitated economic growth by increasing workers' purchasing power.

According to Boyer (1984), industrial relations reflect not just the state of the labour market, but also a compromise which represents the relative strength of workers and employers. Thus the specific features of economic growth in France were only possible because of the nature of that compromise, which he calls the 'wage relationship' (rapport salarial). It consisted mainly of bargaining on productivity increases in exchange for more wages, an absolute novelty compared to the pre-war period, basically linked to important technical changes in many sectors. Workers' demands and collective bargaining concentrated on wages:

> direct wages, including a share of productivity increases and a move towards automatic adjustments to changes in the cost of living;
> fringe benefits, which represent a continuously increasing share of total income (from 1 per cent in 1913 to almost one third of the present income of manual workers).

Thus wage increases strengthened the dynamic of economic accumulation, stimulating mass consumption, and therefore production. The Left tried to carry out the same policy in 1981, but international economic conditions having totally changed within 30 years, it was bound to fail. But during these 30 years, i.e. until about 1980, the employers also used wage increases to prevent workers and unions from questioning their economic policies: wages had become a substitute for qualitative demands.

Wages may therefore be analysed as the only real basis for bargaining, the one which alone allowed the industrial relations system to function. Thus a strike may have been caused by a problem of work organization or job classification, but it could only be resolved by a wage increase. For example, in many plants the introduction of shift work was made possible only by bargaining on wages. In some way or other, that type of bargaining would hide problems of economic policy or power: as they turned these claims into wage increases, the employers succeeded in preventing any embarrassing conflict over their power. Therefore economic growth, because it made it possible to bargain only on wages, added its effects to the political situation (uninterrupted rule by the Right between 1958 and 1981) to prevent the unions from putting forward more fundamental claims. At the same time incomes did indeed increase.

The fact that wages were, for over 30 years, a general substitute for any other demands, is important for analysis of the present position of the French trade unions: it is precisely when economic conditions change, and when wages progressively cease to play their role as substitute, that a 'crisis' of trade unionism begins. But what generates a crisis is much more the content and definition of what can be bargained than the conditions of bargaining.

This argument can only be fully understood after a short description of the two complementary levels of bargaining in France, and of the part played by the state in industrial relations.

1. The general framework is laid down by industry-level agreements; but they only set minimum wage rates, usually very close to the SMIC (statutory minimum wage), which has an impact on all minimum rates.

But since the industry minima represent the wages which the weakest firms can afford, they are usually well below average wage rates. This difference varies according to the structure of the industry: huge and heterogeneous, like the metal or chemical industries; or homogeneous like textiles or petroleum refineries, the first with low, the second with high average wages. Trade union strength also of course has an influence.

2. Company agreements are much more specific about wages, work qualifications, collective rights.

However, there was no legal obligation to bargain collectively either at industry or company level until the legislation of 1982. This explains the many gaps in the system of collective agreements in France. Until the mid-1960s, individual firms like Renault often made the running in the negotiation of both company and industry agreements. Once ratified, such agreements would be extended to all workers in an industry by law. Since 1966 there have been no new agreements of this kind, and most previous agreements have effectively lapsed. Instead, there were important increases in minimum wages and conditions achieved by negotiation or statutory action; but few *general* increases in wages were achieved.

The same period also saw a more precise division of roles between state and entrepreneurs, between social expenditure and investment, a separation defined explicitly for the first time by the CNPF (employers' association) in a document issued in 1965.

The dual bargaining system, together with very strong state intervention, established an industrial relations system based both on the institutionalization of minimum wage rates and on union–employer negotiations strongly conditioned by the particular economic and social conditions of each industry and enterprise. The importance of this differentiation is increasing, especially with diversified technical development.

One of the most recent illustrations of the dual bargaining system is the first agreement on the introduction of new technology, signed in June 1986 between the Banking Employers' Association and all unions except the CGT.[1] The agreement was signed at national level, following discussions which lasted more than two years. It only prescribes the general principles to be taken into account, and included in any further agreement, at company level: working conditions and ergonomics, new employment structures, labour mobility. It particularly emphasizes the need for more effective and continuous vocational training, in order to ensure a more qualified workforce. It also specifies that workers' representatives must be consulted before the implementation of new technology, but after the decision on their introduction has already been taken. A yearly meeting of the national joint committee for banking is to monitor the operation of the new agreement.

In France, the state plays many different roles. Its well-known interventionism is used by unions as a basis and a spring-board for action: social laws often extend to all workers the results obtained in specific sectors, or seek to encourage negotiations between 'social partners'. This latter aspect of state intervention, particularly obvious since 1981, may be considered a means of locking workers and unions into a legal network. But for all governments, the unions are also a counterweight to the impact of big business which rules the economic system. However, it may be said that France lacks those general mechanisms of negotiation which Pizzorno has called 'political exchange' (1978). Such political exchange would require a unified and centralized labour movement, controlled by powerful unions, with the will and power to conform to the rules of the system. But that situation, which appears typical of West Germany, is not found in France. However, institutionalization has been increasing for 15 to 20 years, going beyond just wage bargaining, from the 1968 law which gave unions an official status within establishments to the 'Auroux laws' (named after the Minister of Labour from 1981 to 1984) of 1982. However, weak institutionalization also has some advantages for trade union action, since it makes it easier to mobilize workers, at least when economic conditions are favourable.

But one of the elements which help understand the present situation is that the unions have little institutional basis to face the changing economic and technical structures which transform the conditions of confrontation and bargaining:

> Union membership is weak and has been declining; but there is relative stability at the periphery of those who vote for unions in elections for works committees and, more generally, follow unions in action (or even anticipate them). The impact of these outer circles on union action should not be underestimated: if they do not actually pay subscriptions, they nevertheless have to be considered part of the unions' strength.
>
> The effective protection of workers' representatives in firms remains weak and, above all, very unequal from firm to firm, in spite of a notable improvement since the December 1968 law which gives unions as such a role in the workplace.
>
> The dual bargaining system subordinates the bargaining process to the local economic situation of the company and the 'goodwill' of the employer.

Unions therefore have few institutional weapons with which to face significant changes in employment and in the content of work itself, brought about both by new technologies and structural economic changes which are hard to disentangle. The internal organization of trade unions is less and less in keeping with the organization of labour markets: the national organizations are almost everywhere stronger than their local sections, while it seems that most important confrontations and bargaining, especially on employment, might from now on take place at regional or local level.

Evolution Since the 1970s

Since the beginning of the 1970s, state intervention has aimed both at facilitating structural change in industry and at preventing or correcting adverse social effects. Its intervention in industrial relations has also consisted, well beyond the period of economic growth, in institutionalizing wage increases. But at the same time the state progressively retreated, until 1981, from direct financial investment in industry, particularly in the nationalized sectors. This apparently ambiguous position – more intervention in industrial relations together with less investment – is in itself functional and aims at protecting the victims of structural and technical changes. But another specific function of the state is to foster social consensus. The bargaining relationship of the state with its own wage earners has this function, since the aim is to set an example of willingness to negotiate, especially on wages. But it is also clear that government policy gives employers even more freedom to introduce changes in industry without negotiating with workers and unions. The number of unemployed, which rose steadily but still slowly from 1963, was not yet very high in the early 1970s (under 6 per cent); unemployment benefits were substantial and economic growth was still sufficient to provide new jobs. In short, wages still operated as a general substitute for all claims until the beginning of the 1980s.

Under such circumstances, it is hardly a paradox to say that while economic growth was still predominant elsewhere, it was only in declining companies and industries that workers and unions found it possible to challenge the economic strategies of the employers. Very long sit-down strikes, with popular and political involvement at local level (the LIP model)[2] put forward claims to a right to work, at this precise workplace, claims which increasingly conflicted with technical and structural changes, but could still be supported with some hope of limited success, precisely because they were isolated cases, with new methods of action and strong local support (Casassus and Erbès-Seguin, 1979). But the number of such strikes and their symbolic impact among workers raised clearly – and early – a question which would become predominant in the mid-1980s: the need for trade unions to shift their bargaining strategies from wages to employment.

It should also be noted that French entrepreneurs totally shifted their employment strategies between 1972 and 1975 (Freyssinet, 1982). Until that period most companies, big and small, operating in a restricted labour market with little mobility, tried to retain a permanent workforce. But by 1975 a turning point had been reached with a slow decline in permanent employment, and an equally slow rise in precarious work. The limited mobility of the French workforce, compared to several other Western countries, made even more significant a change which, by American or even West German standards, was very slow. The trend seemed more dramatic because of the parallel, but faster, rise in unemployment due both to structural and technical changes and to a rapidly increasing workforce (post-war baby-boom). The economic 'crisis' which began in the mid-1970s was the efffect of deep internal changes in the economic structure even more than of oil prices. Productivity increases based on new technology slowed down, new investment declined, while harder work was demanded from the workforce. What is now required is a less stable workforce and an increased diversity of wages and employment conditions to meet technical – and often simultaneously geographical – changes. More than an economic crisis, we are now experiencing a long-term shift in the socio-economic conditions of industrial relations.

But at the same time it also became necessary to involve employees in their work, in order to increase the product quality and diminish waste. Quality circles have extended very rapidly in the 1980s: hardly 500 in 1981, over 10,000 in 1984 (Groux and Lévy, 1985). This trend is quite consistent with the efforts of employers to prevent workers and unions from interfering with the decision-making process, since these circles are only concerned with the enforcement of technical standards and aim at making better use of workers' capacities and creativity. Their appeal to workers is probably that they partly answer workers' longing for more control over their work. But it should also be said that the two main trade unions (CGT and CFDT),[3] stimulated by new technology, in particular the computerization process, began some years ago to shift towards less defensive strategies and to seek more participation at company level (Linhart and Linhart, 1985).

The election of the Left in 1981 brought a strong revival of state intervention, both in industry and in social affairs. The nationalization programme answered very deep aspirations among workers and unions: it had been central to their plans for radical economic change, above all for the CGT. But its role remains partly symbolic, since it is well known that nationalization is not sufficient to transform national and international economic conditions.

The government tried above all to urge employers and unions to negotiate:

The December 1981 law which reduced weekly working hours from 40 to 39 was important because unsuccessful negotiations on this issue had continued since 1978, and also because its application required industry-level negotiation.

The October 1982 law established compulsory bargaining once a year on wages, once every 5 years on work classifications at industry level, as well as annual negotiations in each company. Workers' representation and bargaining were also reinforced in very small companies (under 11 workers) which have no delegate or shop steward.

Compulsory bargaining is an important novelty in the industrial relations system. The Auroux laws were explicitly designed to reduce social conflict at shopfloor level, and to meet a long-established union claim through the creation of free expression groups (GE: groupes d'expression directe) at firm level (the August 1982 law, for an experimental period, confirmed and extended to all firms, including the civil service, by the January 1986 law). The aim is to allow direct and collective expression by all workers (but not through unions as such) on 'the content and organization of their work and as the definition and implementation of measures conducive to better working conditions in the company'. These groups were to be established, when possible, through agreements between employers and workers' representatives (trade unions or workers' delegates); or failing that, through the decision of employers, at workplace or company level. The reform is therefore both ambitious and relatively limited. Auroux insisted during the discussion of the bill in Parliament that the aim was not 'an institutional reform of companies' which had been attempted without success several times. 'We want', Auroux said, 'to open new scope for freedom and democracy consistent with the company objectives, because this law aims at effecting changes in work relations'.

The law also extended the role of works councils (comités d'entreprise) which have to be 'consulted' before any important technical change is introduced which might alter employment, vocational qualification, wages or work conditions. However, it does not add much to their actual power, although most of those elected to works councils are trade union representatives; decision-making remains in the hands of the employers, who are only obliged to consult and not negotiate with the works councils.

Results of the First Year's Operation of Free Expression Groups

Free expression groups (Groupes d'Expression: GEs), have been the subject of considerable research, most still incomplete.[4] But the initial findings are consistent, and may be summarized as follows:

1. There are still many open questions as to how GEs may affect trends towards industrial or economic democracy. So far, there has been nothing at company level to compare with British workshop organization. According to several authors, GEs are in no way a step towards industrial democracy, either because of workers' apathy or because of the difficult economic situation (Segrestin in AFERP, 1985), or because GEs simply do not seem important for unions in some industries.[5]

However, in the words of a representative of the Ministry of Labour (in AFERP) they 'encouraged a renewal of interdependence relationships'. Some authors even consider that GEs and more general changes introduced in the industrial relations system at workshop level by the Auroux laws, are potentially decisive steps towards more economic democracy. On this view they can help the workforce to become more active in the introduction of technical change, a development that can only be fostered by more decentralized negotiation (Jeammaud and Lyon-Caen, 1986). Even workers who do not question the decision-making power of the employers seem to become aware of the power gained from simply being able to speak publicly at work. Experience in GEs might thus generate a 'political' consciousness, in the broadest sense of the word, of company problems.

2. Unlike other social laws such as the December 1968 law on shopfloor union representation (which took 10 years to implement in some firms), the GE law was relatively rapidly enforced: 3,000 agreements, representing 45 per cent of the firms after one year, at present over 4,000. Very rarely were GEs established by decision of the employer alone (3.9 per cent according to the enquiry by the journal *Paroles* enquiry).

In practice the implementation was less satisfactory than this may imply. A report by the Ministry of Labour before the final legislation of January 1986 noted that there had been few volunteers to assume the leadership of GEs, and sessions were often poorly planned. In many cases, both manual and staff workers displayed growing enthusiasm, often demanding more training of those who accepted responsibilities in GEs and more time for meetings. But in some cases, the initial enthusiasm faded when the meetings brought no concrete results.

3. Some aspects of shopfloor work relations began to change. An early and predictable result was to make lower and middle management (*cadres*) behave in a less authoritarian and more persuasive manner. But this conclusion must be qualified: some studies (Linhart and Linhart, 1985) show that GEs facilitated better work relations when there was already receptivity to change, but otherwise had little effect.

Unlike West German managements, French employers show very

little interest in industrial relations at company level (Sellier in AFERP). However, an enquiry among employers in November 1985, showed that 64 per cent of them thought GEs were useful and should be maintained.

4. The topics covered by GEs are considered far too narrow by both employers and workers: usually, the agreements reproduce the wording of the law, no more (Ministry of Labour in AFERP; 95 per cent of the agreements, according to the *Paroles* enquiry). Management responses to the claims put forward by GEs vary considerably in content and the time taken to answer them.

5. The most obvious result, on which all researchers agree, is the extreme diversity of the ways in which the agreements have operated. There are many reasons for this. According to Segrestin (AFERP), the most important influences are internal to the company: plant size, previous industrial relations experience, management structure, strength of union organization.

6. The relationship with the other representative institutions at firm level is complementary rather than conflictual. One of the 'Auroux laws' increased the influence of works councils on the introduction of technical change. The creation of GEs usually resulted in co-operative relations with these councils, in particular the Health and Safety Committee (CHS: comité d'hygiène et sécurité), which often acts also as a vehicle for GE proposals. For many years prior to the Auroux laws, the trade unions had been examining their policy towards the introduction of new technology, and regarded the CHS as particularly effective representative institutions (CFDT, 1978; Piotet, 1984; Duchesne, 1984). They had issued proposals which had an important impact on the objectives and wording of the new laws. Collaboration between GE and CHS can be expected to develop further.

Quality circles also seem to be complementary to the GE. In some cases, GE even act as quality circles, but the reverse may also be true.

7. Can the GE be expected to play an important part in the introduction of new technology at plant level? As noted above, the functioning of GEs varies from plant to plant, and that makes it difficult to generalize about their role. However, several trends may be indicated:

GEs often enable workers to develop a new capacity to solve technical problems without any intervention by middle or even senior management. But many problems require the intervention of experts outside the workgroup. GEs might therefore become a forum for organizing the relationship with these outside experts (Martin, 1986). But while the introduction of new technology may enhance the role of the GE in problem-solving, the way in which they operate (who is in charge, how often they meet, on which topics, etc.), and the level at which they were introduced (plant or company) will inevitably affect their role in the introduction of new technology.

Whenever the discussion and resolution of technical problems transcend the workshop, GEs are only one of the places of discussion and debate. It is therefore likely that such problems as the introduction of new technology will be handled primarily in the works council or the general bargaining system.

GEs are explicitly authorized to discuss only questions of work organization. According to the letter of the law, they should not be concerned with general problems, such as the introduction of new technology; otherwise they might interfere with existing institutions such as the works council, personnel delegates or unions. Research confirms that GEs in fact deal very little with such broader questions (Bunel and Bonafé-Schmitt, 1985; Chouraqui *et al.*, 1986).

GEs often include members of middle management, or are organized with their help, or even headed by one of them. They may therefore become channels for downward communications to facilitate management's own strategy for improving productivity. This may help explain why, in a number of companies, GEs function effectively as quality circles.

Overall, then, GEs are unlikely to have a major role in the introduction of new technology, except to assist the matching of work organization and technology.

8. Before the law took effect, the unions' view was far from unanimous. At first, only CFDT was in favour; an internal conference of that union (February 1986) showed much interest on the part of rank and file members. It was rapidly joined by CGC (Confédération Générale des Cadres)[6] after some disagreement concerning the role of middle management in GEs. The other main unions (CGT and Force Ouvrière)[7] were – and in some cases still are – reluctant, and regarded GEs as potential rivals. Some CGT leaders even considered that they had been invented in order to compete with unions at workplace level. And indeed, the law does *not* give unions a central role in establishing GEs; they are just one element among others, although they very often sign the agreement on behalf of the workers.

Do the unions in fact benefit from GEs? According to Linhart (1985) they gained little in companies in which regular discussions with the workforce took place before GEs were introduced. In such cases middle management, not the unions, took control.

But in smaller or middle-sized firms, GE changed authority relations and gave unions the opportunity to increase their impact. However, the concrete results are still limited, and the groups operate in kinds of by-ways, far from the centres of decision-making.

Conclusion

After raising much passion, the expression groups are now an accepted part of the industrial relations system. But the general situation is rather awkward for the unions, and not only for political reasons. It is obvious that the 'Auròux laws' further institutionalized collective bargaining, and generally speaking strengthened the unions at company level. But there are conflicting opinions within each union on the precise consequences.

Above all, the important economic and technical changes of the last decade, and the shattering of previously clear-cut notions such as 'company', 'employment', 'employer', have transformed the basis of what may be bargained. A system no longer exists in which institutionalized wage bargaining is necessary for economic stability. A substitute has to be found, and can only be defined around the conditions of employment for all. But how is still unclear. Meanwhile, all the bargaining machinery of the unions remains centred on wage negotiations. That is one reason why the unions are presently in a difficult position.

Notes

1. The communist-led CGT (Confédération Générale du Travail) is the largest French trade union federation.
2. The LIP watch factory was occupied by the workforce when threatened with closure in 1973, and production continued for several years.
3. The CFDT (Confédération Française Démocratique du Travail), the second main French union, was originally Catholic but shifted to religious neutrality and a socialist political orientation. It has long favoured workers' involvement in decision-making.
4. These are listed in the references under 'Research on Groupes d'Expression'. For an English-language survey see Eyraud and Tchobanian, 1985.
5. Thus the GEs in the building industry merely compiled a list of problems to be discussed but took no further action (AFERP, 1985).
6. The union of managerial staff.
7. The third main confederation, social democrat in orientation.

References

Boyer, R. 1984. 'Wage Labour, Capital Accumulation and the Crisis 1968–82'. *The French Workers' Movement*. Ed. M. Kesselman and G. Groux. London: Allen & Unwin. 17–38.

Casassus, C., and S. Erbès-Seguin. 1974. *L'Intervention Judiciaire et l'Emploi*. Paris: Documentation Française.

CFDT. 1978. *Les Dégâts du Progrès*. Paris: Seuil.

Duchesne, F. 1984. 'La CGT, les Salariés et les Nouvelles Technologies', *Sociologie du Travail*. No. 4, 541–7.

Eyraud, F., and R. Tchobanian. 1985. 'The Auroux Reforms and Company Level Industrial Relations in France'. *British Journal of Industrial Relations*. Vol. 23, no. 2, July, 241–59.

Freyssinet, J. 1982. *Politique d'Emploi des Grands Groupes Industriels*. Grenoble: PUG.

Groux, G., and C. Lévy. 1985. 'Mobilisation Collective et Productivité Economique'. *Revue Française de Sociologie*. XXVI.

Jeammaud, A., and A. Lyon-Caen. 1986. *Droit du Travail, Democratie et Crise*. Actes Sud.

Kesselman, M., and G. Groux (ed.) 1984. *The French Workers' Movement*. London: Allen & Unwin.

Piotet, F. 1984. 'Nouvelles Technologies, Nouveaux Droits. Positions CFDT. *Sociologie du Travail*. No. 4, 535–40.

Pizzorno, A. 1978. *in The Resurgence of Class Conflict in Western Europe*. London: Macmillan. Vol. 2.

Research on Groupes d'Expression

Association française d'étude des relations professionnelles (AFERP). 1985. colloquium. Paris, mimeo. 29 March.

Bernoux, Ph. 1985. *De l'Expression à la Négociation*. Lyon: GLYSI, mimeo. November.

Borzeix, A., D. Linhart, and D. Segrestin. 1984. 'L'Expression Directe et Collective des Salaries, Premier Bilan'. *Politique Aujourd'hui*.

Bunel, J., and J. P. Bonafé-Schmitt. 1985. *Le Triangle de l'Entreprise*. Lyon: GLYSI, mimeo.

Chouraqui, A., and R. Tchobanian. 1986. 'The Employees' Right of Expression Within French Firms'. 7th World Congress of the International Association of Industrial Relations. Hamburg, 1–4 September.

——, A., A. M. Gautier, and R. Tchobanian. 1986. *Premiers Résultats de la Banque de Données PAROLES*. Aix-en Provence: LEST, mimeo.

Linhart, D. 1985. *Revue Française des Affairs Sociales*. no. 2.

Linhart, D., and R. Linhart. 1985. 'Sur la Participation des Travailleurs'. *Politique Aujourd'hui*. October–December.

Martin, D. 1986. 'L'expression des salariés: technique de management ou nouvelle institution?' *Sociologie du Travail*. no. 2.

Les Nouveaux Droits des Travailleurs. 1983. Paris: La Découverte/le Monde.

18

Bargaining over New Technology: A Comparison of France and West Germany

Michèle Tallard

This chapter adopts a comparative perspective, with a focus on France and West Germany. It examines the legal framework and the content of technology agreements in order to identify factors which influence the nature of bargaining over new technology (NT). Two central issues are explored: the ways in which NT introduces new features to collective bargaining; and whether its character can be linked to the distinctive organization of the 'qualification space' in each country.[1]

The Opportunities Created by Legal Enactments

The rights assigned by law in any country to employees and their collective representatives reflect policy decisions on a number of questions:

the type of power conferred: consultation or negotiation rights;
who acquires the rights: trade union representatives or a works council;
the level of involvement: national, regional, local, establishment, workshop;
the stage at which consultation over NT is to occur: the initial planning process, the implementation phase, or monitoring the consequences after the decisions have been taken.
the subjects on which workers may be represented in the decision-making process: investment, employment issues, working conditions.

The following sections outline the legal provisions in each country. In neither case does legislation relate specifically to negotiation over NT.[2] The relevant provisions are those which define in general terms the status of collective agreements and the powers of works councils or works committees.

West Germany: Extensive Legal Powers?

Negotiations occur within the normal framework of industry-wide or regional collective bargaining, or else fall within the co-determination rights of the works council (*Betriebsrat*: BR) as defined by an Act of 1972. Within the German industrial relations system these two processes are formally unrelated. The BR, which has a largely consultative role, is elected by workers within the plant, irrespective of union membership. Its members are subject to a legal 'peace obligation', and are thus prohibited from organizing sanctions in support of their demands. The trade unions negotiate with employers' associations at industry-wide or regional level, reaching legally binding agreements which preclude strikes during their period of currency. Formally the unions have no role within the plant, though members may elect *Vertrauensleute* (literally 'trust persons', or quasi-shop stewards) who may have some influence on the BR.

Industry agreements may contain nothing which could infringe managerial prerogatives, but they may deal with the social consequences of rationalization through measures such as redundancy payments, special provisions for older workers, protection against down-grading, and improvements in working conditions. As will be seen in a later section, the right to strike when national agreements are renegotiated has been used to win important gains on such issues in the printing and metal industries since 1978, and more recently in banking and insurance.

However the main opportunity to negotiate over NT is at enterprise level. Under the 1976 Act the minority employee representation on supervisory boards was increased in firms with over 2,000 employees, theoretically providing parity (as already existed in mining and iron and steel). In practice, however, the employers retain the decisive vote. Thus workers can express a view on investment decisions but cannot override management. Workers' influence is therefore exerted principally within the BR, which has the right to intervene at a number of levels on issues relating to NT:

It must be informed and consulted on any change in technology 'at the appropriate time', and has the right to be advised by an outside expert.

Whenever dismissals are caused by rationalization measures, the employer must submit to the BR a 'social plan' covering training, regrading, and protection for older workers.

The BR has co-determination rights over 'the adoption and implementation of measures designed to control the behaviour and productivity of employees' and over 'changes in job designations which do not accord with current knowledge on labour questions'. On both issues the BR can challenge the proposed measures and seek arbitration from the labour court.

However, while the powers of the BR appear considerable, in practice they are subject to three important restrictions:

1. The notion of 'appropriate time' is very imprecise; thus often the employer can notify the BR very late in the decision-making process. IG Metall[3] estimates that in 60 per cent of cases notification is too late for effective action to be taken.

2. The powers of the BR are limited to the rejection of measures proved contrary to ergonomic principles. But such principles can be established only after lengthy testing – which can take up to eight years, according to union sources; and as Price and Steininger (1987) indicate, there is no *status quo* provision.

Hence the employer can proceed with technological innovation even if a challenge has been submitted within the judicial procedure. Even this limited power has been further restricted by a recent labour court judgment, which ruled that co-determination rights do not exist on subjects covered by legislative provisions; since health and safety are legally regulated, the BR has no co-determination rights, and cannot therefore seek to prescribe medical examinations for VDU operators or to limit their periods of work (Teyssier, 1984).

3. Even before this decision of the labour court, the German employers' organization had included NT in its *Tabu-Katalog*: the list of topics on which member firms were instructed not to negotiate with trade unions. In consequence, the DGB (German TUC) sought an extension of the statutory rights of the BR.

France: Recent Extensions in Workers' Rights at Enterprise Level

The French industrial relations system is characterized by the existence of competing trade union confederations (with distinctive ideological affiliations) and their industry-level subsidiaries. At workplace level exist statutory enterprise committees (comités d'entreprise: CE) with rights to information and consultation. In general, French labour law imposes less procedural constraints on either employers or worker representatives than in Germany.

Workers' participation is covered in the 'Auroux law' of 28 October 1982, which extended the powers of the CEs and the health and safety committees (comités d'hygiène et sécurité: CHS). Previously the CE received indirect information on changes in technology through its regular meetings on employment and working conditions, or through the prescribed notification of rationalization plans involving redundancy. Since 1982 employers have been obliged to consult the CE before any important technological change which could affect employment, skill requirements, earnings, training or working conditions. In establishments with over 300 employees the CE may be advised by an outside expert.

In theory, then, there is considerable scope to intervene, since the employer must provide adequate documentation one month in advance and must justify any refusal to follow the advice of the CE. But in practice there are a number of restrictions. Several judicial rulings demonstrate a reluctance to accept the involvement of worker representatives in decision-making, even in a consultative role:

What constitutes an 'important' change is a matter of dispute.

There is also disagreement on the definition of 'new' technology. Is it any technical innovation, or the introduction of a technological system never previously used in the enterprise? Recent judgments have opted for the second meaning, and have ruled that new applications of existing techniques cannot be regarded as NT.

The burden of proof that a change will affect employment etc., rests with the CE.

In most cases an expert called by the CE is regarded as biased towards the union position.

The CHS also must be consulted before any alteration in job content, work speeds or production standards. It too can call on expert advice, but again the problem of presumption of bias arises.

Another of the 'Auroux laws', the annual requirement to negotiate the length and organization of work time, covers the introduction of NT to the extent that this creates changes in this area. One further 'Auroux law' must also be mentioned: workers' 'right of expression' on work-related issues.[4] During the parliamentary debate on this statute it was emphasized that this would permit groups of workers to discuss the implementation of NT; but in the years when it has first been in force it has mainly been applied to more general working conditions. But the extension of the law to production issues should increase the scope for 'expression groups' to intervene on questions of NT.

In 1984 no agreement on issues of work flexibility was reached in the national negotiations between the central trade union and employer organizations; there is therefore no contractual obligation to negotiate on NT at industry level.[5] However, such a requirement may derive from provisions in the Auroux legislation: one requiring negotiators to consider job classifications every five years; another prescribing industry-level negotiations on training objectives and methods; and finally a law covering redeployment of workers affected by rationalization and specifying periods of retraining.

This brief survey of French legislation affecting NT shows that the principal effect of the 'Auroux laws' has been to strengthen employee representation at enterprise level. 'Rather than strengthening collective bargaining, these provisions bring a greater involvement of employees in decision-making' (Lemaitre *et al.*, 1986).[6]

Franco–German Convergence

It is possible to identify a process of convergence between the two countries in bargaining over NT, with a growing influence of workplace committees (though their form is very different in each country). In both cases, however, these bodies face serious obstacles in seeking to influence management decisions, as can be seen by the number of cases in both countries where the law has been tested in the courts.

One issue to consider when examining NT and collective bargaining is whether distinct technology agreements on the British model exist, or whether NT is covered within more general agreements on rationalization, training and job classification. In other words, do unions approach NT like any other bargaining issue, or do they have to adopt new strategies and tactics? The latter view was expressed by the French industrial relations academic J. D. Reynaud at a trade union conference in 1980: negotiating over NT is only 'partially the same' as ordinary collective bargaining: 'it involves creating new methods and new procedures'.

West Germany: a Three-Stage Evolution

Within German industrial relations there has long been consensus on the need for productivity and hence the inevitability of innovation. The employer's function is to ensure economic growth, that of the unions to moderate the impact of rationalization on the workforce, thus ensuring that the process of change is socially acceptable.

These principles underlay the first stage in the evolution of the German approach to NT, beginning in the 1970s. The central aim was to improve working conditions and protect those categories of workers who were seen as particularly vulnerable to redundancy.

Since 1967, agreements in the metal industry have required social plans accompanying rationalization to include provisions for retraining and redeployment. In 1973 an agreement in the North Baden-Württemberg region[7] provided guarantees against downgrading and defined other principles to govern rationalization plans. What some commentators have called the 'optimistic phase' reached its peak in the mid-1970s with the participation of the DGB in the 'humanization of work' (*Humanisierung der Arbeit*: HdA) programme initiated by the government.[8] Many industry-level agreements included clauses on working conditions, particularly for VDU operators, as did over a hundred plant agreements between managements and BR.

At the end of this period a more defensive strategy was pursued in some sectors of the trade union movement, in particular in the engineering and printing industries in Baden-Württemberg. Two points require emphasis:

> The main basis for formal union intervention on NT issues is in the industry-level collective bargaining which occurs at fixed intervals. Here the main focus of negotiation is the impact on the *status* of employees of technological change, for example the effects on job classifications and wage levels (Tallard, 1984).
> The BR has the right to monitor and constrain the consequences of NT in order to protect employees, but not to initiate proposals on its introduction.

Two important elaborations of these principles were the 1978 agreements in printing and engineering in Baden-Württemberg: the first dealing specifically with NT, the second including the topic within the regular general agreement. Both followed long and bitter strikes and lockouts, and sought to protect the position of skilled workers through guarantees of job security for older workers and prohibition of down-grading. Clauses were also included on health and safety, particularly in respect to VDU operators. Such agreements spread to other regions and now cover roughly 900,000 metalworkers. More than half of all industrial workers are now subject to agreements providing protection for older workers. More recently, job protection agreements have been signed in banking and insurance.

A third phase commenced at the end of the 1970s, with the BR playing an enhanced role. Under the auspices of the HdA programme the DGB received funds in 1979 to establish technology advice centres with experts able to assist BRs faced by NT. The aim was to analyse employers' proposals and also possibly to formulate alternative plans which would take account of concerns with job security and in a broader sense accommodate NT to workers' interests. Not all the cases taken up have resulted in agreements; but the aim of the centres has also been to help develop a set of criteria which a project should satisfy in order to be considered favourable to workers' interests. Thus a kind of check-list has been produced which any BR can apply.

However a recent IG Metall survey of BRs shows that this strategy has achieved little. Most plant agreements on NT covered redeployment, job security, job classifications and wage levels. But very few covered co-determination rights or such aspects of work organization as shift-work. In more than a third of all cases there was no negotiation at all when NT was introduced. The overall picture was of negotiation taking place only *after* management has made the decision. Could the situation have been different? Most cases submitted to arbitration arise when a BR attempts to interfere in managerial prerogatives.

In any event, even if the recent change in union strategy has had only a marginal effect, it is significant for two reasons. It demonstrates a new awareness of the importance of workplace negotiation – previously considered a secondary element in German industrial relations because the unions are not formally involved. The special nature of bargaining over NT means that workplace negotiation must be accepted and co-ordinated. Unions are obliged to increase their involvement at this level in order to prevent the 'consensual incorpora-tion' of BRs within management strategy (Price and Steininger, 1987). The increased importance of workplace negotiation stems also from the national agreement reducing working hours from 40 a week to an average of 38.5; each individual's weekly hours (within a range of 37 to 40) was to be determined at establishment level.

This development creates a major dilemma for German trade unions. 'It is increasingly necessary to intervene at local level, in the factory or workshop where NT actually takes shape. But at this level the workers cannot take action; if there is a dispute they can only ask for arbitration.'[9] The further significance of the new phase of union strategy is thus the shift from the initial optimistic and subsequent defensive approaches to a more radical and offensive stance. This could entail a more political approach to the problem of NT.

France: the Difficulties of Bargaining over NT

Recent comparative studies (e.g. ETUI, 1985) show that collective bargaining over NT, especially at industry level, is not well developed in France. Before 1980 there was no negotiation on the subject outside banking and insurance, although unions had developed strategies on NT from the mid-1970s. At enterprise level, the Auroux legislation seems to have created new scope for negotiation.

At industry level, NT is subsumed within the broader bargaining agenda.[10] Until the May 1986 agreements in banking and the provincial press, there were no binding NT agreements at this level; merely various policy declarations. In 1976 an agreement on working conditions in the insurance industry *encouraged* job enrichment, and one with employers' organizations in the Paris region covered VDU work. But the most important agreement of this kind was reached in 1983 in banking. It establishes *non-obligatory guidelines* encouraging firms to make adequate provision on such matters as maximum work spells, seating arrangements and illumination for VDU operators. While these recommendations are not widely known, they appear to have helped in initiating negotiations in a number of banks.

In many industry agreements, even when NT is not the main subject covered, it has nevertheless been the reason for initiating negotiations on topics such as training, job classification and shorter working hours, for example, the grading agreements in engineering, chemicals and textiles between 1975 and 1980. These alter the grading criteria to take account of new forms of work organization, or create new categories for operators of particular machines. Many agreements were signed following the 1984 legislation requiring industry-level negotiation on training; most emphasized the special training requirements of NT, most notably in chemicals and textiles. The latter was actually entitled 'master agreement on training for NT', and followed an agreement between the employers' organizations and the government, which undertook to fund a programme of NT training.

It is also necessary to mention the December 1984 draft multi-industry agreement which was never ratified, because it included general provisions on flexibility and deregulation which were ultimately rejected by the unions. A specific section dealing with technological change defined procedures for notification and consultation with CEs. Some unions wished to approve this section at least, but the employers would not accept a separate agreement on this issue. Subsequent industry-level agreements have however often used this draft agreement as a point of reference.

In May 1986 two industry agreements on technological change were approved, in the provincial press and banking.[11] The latter is particularly important, since it followed two years of negotiation and elaborates the legal requirements for consultation with representative bodies at the workplace. It is designed as a framework for progress at enterprise level, defining the main rules to be observed. Unlike the 1983 agreement it is not merely declaratory:

It defines the role of the CHS and the various topics on which it should be consulted.

Without defining their content it encourages the provision, and also the rewarding of successful completion of training programmes.

It specifies the rules and methods for involvement of the CE. This should be informed of any NT project before the final decision on implementation; the information should include details of the scheme, its implications for job structure and content, training, and working conditions. It is also laid down that the CE should not challenge managerial prerogative or the principle of introducing NT.

The application of the agreement is to be monitored by a joint committee.

The agreement for provincial daily newspapers is more concerned with job security and preserving pay levels. It prescribes further national negotiations on job classification and shorter working hours.

At enterprise level, agreements on NT are less significant. Understandably, most such agreements are in banking and insurance, mostly negotiated since 1980 and in particular immediately after the 1983 national agreement. They cover maximum work periods and other conditions for VDU operators, and in some cases guarantee job security. The most innovative agreements are in Crédit du Nord, whose CE must be consulted *before* a computerization plan is drawn up, and in Crédit Lyonnais, where an agreement for computer operators includes clauses on mobility, training, medical checks and grading.

In insurance, the most important negotiations have been on job security, VDU work, and the organization of the working week. The CE, the CHS and the 'expression groups' established by the Auroux legislation have all been vehicles for improvements in working conditions.

There have been fewer negotiations in other industries, particularly in manufacturing.[12] Those that have occurred have been confined to the large enterprises and fall within three categories:

rationalization plans, for example an agreement on early retirement and retraining in a large provincial press group;
job classification agreements, primarily in the motor industry, covering grading of jobs on new machines. Similar agreements have been reached in the cement industry;
consultation procedures. Most classification agreements have also involved the development of worker participation mechanisms such as quality circles. These procedures are designed to accommodate workers and their representatives to technological change, but such groups are rarely permitted a say on investment decisions or choice of technology; these areas remain taboo. Participation in the sense of worker influence on decision-making is neither achieved nor intended (Bunel and Saglio, 1985).

Special mention must be made of the discussions in the steel industry following the 1979 rationalization plan. At industry and company level the unions submitted proposals to avoid redundancies; while at establishment or even workplace level the employers experimented with all manner of participation mechanisms. Their aim was to by-pass the unions by involving workers directly in changes to job content and skill structures (Groux and Lévy, 1985).

Such developments, while of growing significance in large firms, remain unusual. A recent survey of a hundred firms with less than 500 employees which had introduced NT found that in 60 per cent there had been no consultation whatever, and in the remainder there had merely been a meeting to inform the CE (d'Iribarne and Fossati, 1986).

In summary, most national agreements on NT seek to define ground-rules for the future development of the industry. They are concerned primarily with employment levels, grading principles, working time, and the role of the CE and the unions in company decision-making. The main problem for the unions is to influence the new vocational and skill structure emerging with the spread of NT (Mercier, 1985).

The growing importance of workplace representatives following the Auroux legislation, and new managerial policies to encourage direct worker participation, will not necessarily lead to an extension of collective agreement. The outcome could be a broader involvement of social actors on the shopfloor: a 'new bargaining system' involving a 'culture of negotiation' (Lasfargue, 1986).

Conclusion

This comparative analysis reveals a gap of a decade between the legislative changes in Germany (the 1972 Co-determination Act) and France (the Auroux laws), and in the development of bargaining over NT in the two countries. This is attributable to the lag in technical change in France.

Several common features also emerge. While industry-level agreements impose stricter guidelines in Germany than in France, in both countries the main focus of negotiation is the enterprise, and a range of procedures have developed for involvement of the BR or CE. Unions in both countries, faced by the complexity of the problems raised by NT, have used their legal rights to call on experts for advice.[13] Involvement is, however, often limited to discussing the *consequences* of NT; scarcely ever are unions involved in investment decisions or the choice of technology. Employers resist this as an infringement of their prerogatives, and sometimes unions themselves do not see this as their function.

Despite these points of similarity, the scope and outcome of NT bargaining in the two countries seem to reflect contrasting logics stemming from the vocational structures of each. In Germany the main national agreements have sought to protect the skilled worker; the cost of deskilling has been made so high as to affect the employer's choice of technology. The German system is designed above all to protect 'job property rights'; 'job classifications, use of skills and pay levels' (Price and Steininger, 1987). Significantly, a study of the main German car companies (Jürgens *et al.*, 1984) showed agreement between management, unions and BR on the principle of NT, but conflict over the adaptation of the job structure to protect 'the status of skilled workers'. Hence the BRs called for the rotation of less skilled tasks within work teams so that individual jobs would remain skilled.

In France the main agreements at both national and company level have concerned grading and training. Because of the limited importance of formally certified qualifications in the French vocational structure, occupational status owes more to the worker's classification within a relatively rigid grading system. A study of the introduction of numerically controlled machine tools (Eyraud *et al.*, 1984) demonstrated the social significance of the grading structure: 'everything is arranged as if firms adapt job allocation to this structure, organizing training accordingly'.

Training is the central focus of NT bargaining: it has been the basis of a national multi-industry agreement and of a law requiring negotiations at industry level. In addition the 'Auroux laws' seek to increase the flexibility of the system by prescribing the joint examination of grading structures every five years. Such negotiations may entail a modification of the French vocational system.

In short, NT bargaining in Germany centres on the flexibility of the main rules of the system; in France, on structural changes to the rules. Negotiation has to satisfy different types of needs in the face of economic crisis. In Germany, because rules are flexible, change can occur organically; but in France, because rules are rigid, they themselves must be altered (Silvestre, 1986).

Notes

1. This term is used by Maurice *et al.* (1984) to denote the distinctive ways in which education and vocational training, the division of labour and its social organization, and the institutions of industrial relations, determine the employment opportunities available to the individual.
2. The one exception is the German data protection legislation of 1976 which covers the collection of personal information on employees.
3. The Metalworkers' Industrial Union, the largest trade union in West Germany.

4. For a more detailed discussion of this law, see chapter 17 by Sabine Erbès-Seguin.
5. Under French labour law, a formal agreement reached by 'representative' organizations is legally binding in all sectors in which they have jurisdiction.
6. In both France and Germany the workplace representative machinery prescribed by law is not formally related to trade union organization. In Germany, where collective bargaining is strongly entrenched, the two are in practice closely linked; this is far less the case in France, where unionism is politically fragmented and collective bargaining only weakly established.
7. The major location of the German motor industry.
8. The DGB nominated a third of the members of a commission which selected projects designed to improve working conditions or evaluate new ergonomic principles.
9. This comment was made in an interview with me some years ago by the German industrial relations expert U. Briefs.
10. Most of the examples discussed in this paragraph are drawn from Sellier (1984).
11. The first was ratified by all the unions, the second by all except the CGT.
12. A first one was signed on 21 January 1987 in the metal industry.
13. This has led in France to the claim that 'the trade union expert has displaced the class warrior' (Groux, 1986).

References

Bunel, J., and J. Saglio. 1985. *La construction du consensus technologique.* GLYSI/EEC. Brussels: 1985.

European Trade Union Institute. 1985. *Technology and Collective Bargaining: a Review of 10 Years of European Experience.* Brussels: ETUI.

Eyraud, F., M. Maurice, A. d'Iribarne, and F. Rychener. 1984. 'Développement des qualifications et apprentissage par l'entreprise des nouvelles technologies: le cas des machines-outils à commande numérique (MOCN) dans l'industrie mécanique'. *Sociologie du Travail.* no. 4.

Groux, G. 1986. Contribution to a trade union day school, 20 March 1986.

——., and C. Lévy. 1985. *Mutations industrielles et nouvelles pratiques sociales: Le cas de la sidérurgie française.* Brussels: CNAM-Laboratoire de sociologie du travail et des relations professionnelles–EEC.

d'Iribarne, A., and H. Fossati. 1986. Diffusion des technologies informatisées de production, emploi et formation dans les petits et moyens établissements de la région PACA. Document LEST 86–2.

Jürgens, U., K. Dohse, and T. Malsch. 1984. *New Production Concepts in West German Car Plants.* Berlin: IIVG, Pre 841223.

Lasfargue, Y. 1986. 'La panne ou la négociation'. Project No. 9.

Lemaitre, A., B. Reynes, and F. Teyssier. 1986. *Droit du travail et automatisation: La France.* CEJEE.

Maurice, M., F. Sellier, and J. Silvestre. 1984. 'The Search for a Societal Effect in the Production of Company Hierarchy: A Comparison of France and Germany'. *Internal Labor Markets.* Ed. P. Osterman. Cambridge, Mass.: MIT Press.

Mercier, N. 1985. *Seconde étape de l'informatisation: une banque.* Laboratoire de sociologie du travail et des relations professionnelles CNAM.

Price, R. and S. Steininger. 1987. 'The Control of New Technologies: Union Strategies in West Germany'. *New Technology, Work and Employment*.

Reynaud, J. D. 1979. 'Conflit du travail et régulation sociale'. *Revue Française de Sociologie*, Vol. IX.

Sellier, F. 1984. 'L'impact des nouvelles technologies sur l'action syndicale et le système français de relations industrielles'. LEST 1984.

Silvestre, J. J. 1986. 'Marchés du travail et crise économique – de la mobilité à la flexibilité'. *Formation-Emploi*, June.

Tallard, M. 1984. 'La prise en compte des nouvelles technologies dans la négociation collective: le cas de la RFA'. *Sociologie du Travail*, no. 4.

Teyssier, F. 1984. *Droit du travail et automatisation: la RFA*. CEJEE.

Index

Fordism, 3, 4, 5, 6, 48, 49, 52, 57, 61, 69,
72, 91, 97, 101, 103, 114, 115, 118,
119, 122
Fossati, H., 293
Fox, A., 3, 32, 34, 35, 51, 91, 200, 212,
257
France, 14–16, 44n.3, 92, 97, 98, 205,
216n.1
industrial relations in, 272–82, 286–8,
291–4
Francis, A., 216n.4, 250
Freeman, C., 193
Frenkel, S.J., 234, 237, 239, 240, 245n.3,
246n.7
Freyssinet, J., 277
Friedman, A., 170, 174
Friedman, D., 102, 114
Friedmann, G., 23, 97
Fröbel, F., 113
Fryer, R.H., 206
'full circuit of capital', 147
full-time trade union official(s), 133
functional flexibility, 55

Galjaard, J.H., 118
Gallino, L., 98
Galtung, J., 191, 192
GEC, 250
General Motors (GM), 6, 105, 106, 108,
109, 111, 112, 116, 117, 118,
124n.6, 221, 224
Absence Control Programme, 108
Fiero plant, 106, 119, 223
Hamtramck Detroit plant, 118
Job Opportunity Bank-Security, 105
Kansas plant, 108
Quality of Work Life Program, 104,
106, 120
Tarrytown plant, 111, 123
general union(s), 175, 180, 181
Gensior, S., 22
Gilchrist, R.R., 144
Goldthorpe, J.H., 34
Goodman, J.D., 153
Gorz, A., 58, 59, 148
Gough, J., 57
government intervention, *see* state
intervention
Granick, D., 151
Grant, W., 185n.3
Gregory, D., 56
Groux, G., 277, 293, 295n.12
growth areas, main, 194–5
Grunberg, L., 98
Guest, R.H., 97, 101
Gulowsen, J., 166
Gunn, T.G., 22

Hales, M. 149
harmonization of employer and worker
interests, 4, 50, 53–5
Harrison, B., 95
Harvard Business Review, 111
Hastings, S. 56
Hawes, W., 256
Hawke, R., 235, 238, 244
Hayashi, M., 115
Head Office Officials Consultative
Group, 133
health and safety, 14, 208, 240, 241, 249,
250, 256, 280, 286, 289
Hecksher, C.C., 230n.7
Hedberg, B., 155
Heritage, J., 131
Heydebrand, W.V., 24
'high flyers', 132, 137
'high-technology cottage industry', 5, 50
high trust relationship(s), 91–2, 138, 200,
266
Hill, R.C., 116
Hillage, J., 249
Hirsch, F., 58
Hirschhorn, L.T., 231n.13
Hirschman, A., 161,198
historical dimension of institutional
change, 4–5, 70–2
Hoff, A., 102
Hoffman, K., 216n.2
Holusha, J., 108
homeworking, 215
Honda, 109, 110
'hot autumn' of 1969, 93
Hotz-Hart, B., 33, 62
hours of work, 132, 134, 136, 138, 139,
140, 191, 207, 234, 249, 278, 290,
291
Huber, J., 195
human resource management, 212, 215,
239
Human Systems Management, 26
Humanisierung der Arbeit, 289
'humanization of working life', 16
Hunter, L.C., 213
Hyman, R., 51, 52, 70, 206, 264

Iacocca, Lee, 124n.6
ICFTU, 216n.2
ICI, 144
'ideal type' union policy, 11, 206–8, 209,
216, 216n.3
IG Metall, 261, 286, 290, 294n.3
Il Sole 24 Ore, 98
Income Data Services, 181
incomes policy, 81
indirect worker(s), 254

Index by Annemarie Flanders